SpringerBriefs in Public Health

SpringerBriefs in Public Health present concise summaries of cutting-edge research and practical applications from across the entire field of public health, with contributions from medicine, bioethics, health economics, public policy, biostatistics, and sociology.

The focus of the series is to highlight current topics in public health of interest to a global audience, including health care policy; social determinants of health; health issues in developing countries; new research methods; chronic and infectious disease epidemics; and innovative health interventions.

Featuring compact volumes of 50 to 125 pages, the series covers a range of content from professional to academic. Possible volumes in the series may consist of timely reports of state-of-the art analytical techniques, reports from the field, snapshots of hot and/or emerging topics, elaborated theses, literature reviews, and in-depth case studies. Both solicited and unsolicited manuscripts are considered for publication in this series.

Briefs are published as part of Springer's eBook collection, with millions of users worldwide. In addition, Briefs are available for individual print and electronic purchase.

Briefs are characterized by fast, global electronic dissemination, standard publishing contracts, easy-to-use manuscript preparation and formatting guidelines, and expedited production schedules. We aim for publication 8–12 weeks after acceptance.

More information about this series at http://www.springer.com/series/10138

Deborah Wallace • Rodrick Wallace

COVID-19 in New York City

An Ecology of Race and Class Oppression

 Springer

J039

BPS3

Deborah Wallace
Independent Disease Ecologist
New York, NY, USA

Rodrick Wallace
Division of Epidemiology
NYS Psychiatric Institute
Columbia University Medical Center
New York, NY, USA

ISSN 2192-3698   ISSN 2192-3701   (electronic)
SpringerBriefs in Public Health
ISBN 978-3-030-59623-1   ISBN 978-3-030-59624-8   (eBook)
https://doi.org/10.1007/978-3-030-59624-8

This Springer imprint is published by the registered company Springer Nature Switzerland AG
The registered company address is: Gewerbestrasse 11, 6330 Cham, Switzerland

2/7/22

Preface

The liberal image of New York City gaslights a huge proportion of the literate world. Massey and Denton (1993) found, however, that New York maintained and deepened racial segregation beyond that of most other major American cities. McCord and Freeman (1990) shocked the public health research community with their detailed comparison of life expectancy of men living in Harlem with white New York men and gave impetus to health disparities research and to the National Institutes of Health (NIH) minorities' health entity that was first a center and now a full institute.

Verso Books published two volumes at nearly the same time: *The Assassination of New York* by Robert Fitch and *A Plague on Your Houses: How New York City Was Burned Down and National Public Health Crumbled* by the authors of this book. Fitch revealed how the City's Master Plans plotted the removal of industry from Manhattan and the removal of blue-collar workers as well. Fitch quoted documents showing that the real estate establishment, the wealthy powers, and the elected officials viewed both industry and blue-collar workers as coarse, dirty, and unworthy of the space coveted by them. We described how the John Lindsay Administration and New York City mayors thereafter used unsound algorithms dreamed up by the Rand Corporation to gut fire service in poor neighborhoods of color, after legal urban renewal became politically unviable (Wallace D, Wallace R, 1998). The 1969 Master Plan required massive destruction of these neighborhoods and did not hide their targeting. Class and race/ethnicity bigoted stereotyping provided a cover for landgrabs as the FIRE sector (Finance, Insurance, and Real Estate) pursued ever-higher land and building values for ever-greater profits.

The booting of industry from large swaths of New York City wounded the local economy and made it unable to offer much resistance to the later deindustrialization that began in the 1960s. The FIRE sector ended up the major economic engine of New York City and far too powerful for a truly democratic municipality. Wall Street rules. This lack of economic diversity also confers a fragile brittleness. Economic ups and downs have much greater impact on New York City than on the rest of the country, with city and state budgets reflecting the crests and troughs. Making working class and poor neighborhoods of color bear the brunt of the troughs without getting any of the "goodies" of the crests has been the standard operating procedure (SOP) for many decades. Even during crests, these neighborhoods suffer deprivation of such essential services as fire control, sanitation, education, transportation, and even non-corrupt policing. The Knapp Commission investigated the corruption of police precincts in the 1960s–1970s and the Mollen Commission, in the 1980s–1990s. In both eras, these precincts served poor areas such as Brownsville, Crown Heights, Harlem, and the South Bronx.

We documented the cascade of disasters that the gutting of fire service to poor densely populated neighborhoods triggered. For full description, see our 1998 book and two peer-reviewed publications (Wallace 2011; Wallace and Wallace 1990). Public health and safety unraveled at the neighborhood, municipality, and metropolitan regional and national levels. Donna Shalala was the executive director of the Municipal Assistance Corporation set up to oversee New York City finances during the so-called budget crisis of 1975 and later. She preached the gospel of "less is more," touted New York as a national laboratory for increasing governmental efficiency, and loved the Rand algorithms. She became deputy commissioner of HUD under President Jimmy Carter and spread the algorithms and others like them nationally because of their "success" in New York City. Large parts of cities burned down and created vulnerability to the public health crisis emanating from New York City. President Bill Clinton named her Secretary of Health and Human Services, where she donned the mask of superhero to deal with the problems that she helped cause earlier. This spurious hero's mask has now been donned by New York Governor Andrew Cuomo, who is dealing with the COVID problems that his policies caused or exacerbated such as inadequate hospital resources.

Our past research explored the spread of diseases such as AIDS and tuberculosis among New York City neighborhoods and among the counties of the New York City metropolitan region. We also looked at the spread of AIDS among the major metropolitan regions. Like AIDS and tuberculosis (TB), COVID-19 hits marginalized communities and population sectors harder than wealthy, white ones. Black and Latinx people, the elderly, and the working class and poor suffer higher incidence of morbidity and mortality from the virus, as well as more severe illness (CDC 2020; National League of Cities 2020).

This book has four major chapters as well as a concluding epilogue. Chapter 1 does not directly delve into COVID-19 epidemiology but focuses on premature mortality rates (deaths below age 65) of the 59 community districts of New York City and the socioeconomic (SE) and public health system, which produce these rates. Neither the City nor the State reports COVID statistics for the community districts. High premature mortality rate may indicate accelerated aging and a wider age range of vulnerability to this novel coronavirus.

Chapter 2 analyzes the ZIP code area geography of the three indices of COVID-19: percentage of positive swabs, case incidence per 100,000, and mortality incidence per 100,000. We compare Brooklyn with Queens and Manhattan with the Bronx. Brooklyn and Queens have similar total populations and are contiguous. Manhattan and Bronx have similar total populations and formed an epidemiological unit during the late 1970s to early 1990s syndemic; they are within walking distance of each other via the bridge over the Harlem River and have several subway and bus connections.

Chapter 3 repeats our work on the spread of AIDS, tuberculosis, violent crime, and low-weight births over the 24 counties of the New York City metropolitan region, but for COVID-19 deaths. The index of contact depends on the journey-to-work data from the US Bureau of the Census, and the index of vulnerability is county poverty rate. A striking contrast is found between infectious and chronic disease, with the Bronx section of New York City, wounded by policies of "planned shrinkage" aimed at dispersing minority voting blocs, most marked by death.

Thus, we will examine three levels of organization: community districts, ZIP code areas, and counties. The SE factors that determine either vulnerability to COVID or the actual COVID indicators may differ from level to level. The roles of race/ethnicity and class may differ between levels of organization. Results from these analyses could prove useful for interventions by public policy, if there is political will. The authorities concerned themselves with "flattening the curve" to mitigate the flood of sick people into the hospitals. Public policies for immediate slowing of the pandemic were quickly developed and enforced. We have not, to date, seen production of long-term policies for prevention of either further waves of COVID-19 or future waves of other diseases that jump from animals to humans.

Chapter 4 discusses the need for "upstream" prevention at the source of these jumps of diseases from animals to humans and what that prevention must entail, focusing on neoliberal land use policies and their impact on pandemic penetrance. One aspect of the analysis examines the impact of stochastic burden—"noise"—on the now-infamous "R_0" deterministic epidemic model.

Chapter 5 summarizes our findings and places them in context.

One word about our data. All data are problematic in all community explorations. Yet, epidemiologists find ways of analyzing them to reveal patterns. The COVID data from the early months of the New York City pandemic suffer from lack of adequate testing for the presence of the virus and for antibodies, the indicators of past infection. Chapters 2 and 3 analyze the patterns of COVID indicators at the ZIP code area and county levels, respectively. Chapter 2 relies on the percentage of positive swab samples and incidence of confirmed COVID cases and deaths, which, in turn, also rely on swab samples.

Testing depends on people presenting for the tests and being accepted as deserving tests. In the early weeks, test gear was in short supply and people were discouraged from seeking tests unless they were very ill. Furthermore, many people were turned away even if they were very ill. We know that hospitalized active-disease cases usually were tested. We do not know whether test criteria were uniformly applied over the neighborhoods for non-hospitalized sick people or worried asymptomatic people. Early maps of cases and deaths indicated skimpy early testing in poor neighborhoods of color: Harlem, the South Bronx, and parts of Brooklyn and Queens. Hospitalizations of severely ill sufferers show a different picture. We pushed the dates of our data back as far as possible because later data reflected wider and more even testing as the poor neighborhoods became obvious community-based disease foci, apart from the nursing homes. Other factors hampering testing and hospitalizations in poor communities include deep distrust of authorities and severe shortage of hospital beds and staff. Most hospitals in poor areas were utterly overwhelmed; some such as Elmhurst Hospital in Queens and SUNY Downstate Medical Center in Brooklyn featured in news stories as examples of the impact of COVID on medical care facilities.

Chapter 3 analyzes COVID deaths at the county level for the crest of the first wave, April 2020. As with the data in Chap. 2, these data depend on testing. Although New York City began reporting both confirmed and probable COVID deaths a bit later, the state reported and reports only confirmed deaths. We intend to examine all deaths from all causes at the county level after the first wave has ended and compare them with all deaths from all causes of several previous years to find the real impact of the COVID pandemic: COVID deaths and elevated deaths from other causes due indirectly to the pandemic.

We believe that the actual incidence of illness and death from COVID is much higher in poor communities than what City and state report. The patterns in Chaps. 2 and 3 are muted compared with the likely reality.

New York, NY, USA   Deborah Wallace
New York, NY, USA   Rodrick Wallace

References

CDC (2020) https://www.cdc.gov/coronavirus/2019-ncov/need-extra-precautions/index.html
Fitch R (1993) The Assassination of New York. Verso, New York and London
Massey D, Denton N (1993) American Apartheid: segregation and the making of the underclass. Harvard University Press, Cambridge
McCord C, Freeman H (1990) Excess mortality in Harlem. N Engl J Med 322:173–177
National League of Cities (2020) https://covid19.nlc.org/resources/resources-by-topic/vulnerable-populations/
Wallace D (2011) Discriminatory mass de-housing and low-weight births: scales of geography, time, and level. J Urban Health 88:454–468
Wallace D, Wallace R (1998) A plague on your houses: how New York City was burned down and National Public Health Crumbled. Verso, New York and London
Wallace R, Wallace D (1990) Origins of public health collapse in New York City: the dynamics of planned shrinkage, contagious urban decay, and social disintegration. Bull NYAM 66:391–434

Contents

About the Authors

Deborah Wallace, PhD, received her PhD in symbiosis ecology from Columbia University in New York City in 1971. In 1972, she became an environmental studies manager at Consolidated Edison Co. and participated in pioneering environmental impact assessment. She became a manager of biological and public health studies at New York State Power Authority in 1974 and remained there until early 1982. In 1980 she completed a MiniResidency in epidemiology at Mount Sinai Medical Center. In the mid-1970s, she also founded Public Interest Scientific Consulting Service, which produced impact assessments of massive cuts in fire service in New York City. She also probed the health threats that plastics in fires posed to firefighters and became an expert witness in litigation for plaintiffs in large fires fueled by plastics. From 1985 to 1991, she worked for Barry Commoner at the Center for the Biology of Natural Systems at Queens College. From 1991 to 2010, she tested consumer products and services for their environmental and health impacts at Consumers Union. She retired in 2010 but continues data analysis, research, and scientific publications. Her first paper was published in 1975, and her last publication, a book, in 2019.

Rodrick Wallace, PhD, is a research scientist in the Division of Epidemiology at the New York State Psychiatric Institute, affiliated with Columbia University's Department of Psychiatry in New York City. He has an undergraduate degree in mathematics and a PhD in physics from Columbia University, and completed post-doctoral training in the epidemiology of mental disorders at Rutgers University in New Brunswick, New Jersey. He worked as a public interest lobbyist, including two decades conducting empirical studies of fire service deployment, and subsequently received an Investigator Award in Health Policy Research from the Robert Wood Johnson Foundation. In addition to material on public health and public policy, he has published peer-reviewed studies modeling evolutionary process and heterodox economics, as well as many quantitative analyses of institutional and machine cognition. He publishes in the military science literature and, in 2019, received one of the UK MoD RUSI Trench Gascoigne Essay Awards.

Chapter 1
Premature Death Rate Geography in New York City: Implications for COVID-19

1.1 Introduction

As of May 9, 2020 (the date of this writing), New York State (NYS) had the highest number of COVID-19 cases and fatalities of the 50 states (CDC 2020); New York City (NYC), the great majority of those cases and fatalities (NYS 2020). A study by Yale's School of Public Health and reported in *The New York Times* (Carey and Glanz 2020) demonstrated that travel patterns had spread NYC's particular genetic COVID fingerprint around the country. We had seen that New York City was the national epicenter for AIDS and spread AIDS through the network of metropolitan regions (Wallace et al. 1997). Deliberate destruction of neighborhoods of color by public policy set New York City on that AIDS (and tuberculosis [TB], low-weight birth, violence) trajectory (Wallace and Wallace 1998). The public health and socioeconomic system that provides the background for the COVID epicenter in New York City merits examination.

Years ago, the now-deceased Center for Urban Epidemiological Studies (New York Academy of Medicine) explored year 2000 HIV deaths across the 59 community districts (CDs) of New York, mortality rates from other causes, and associations with socioeconomic and urban environmental factors (Wallace et al. 2016).

HIV/AIDS mortality rate showed the highest dispersion of health outcomes: maximum 110 times minimum. Homicide, liver mortality, and drug mortality associated in multivariate regression with HIV/AIDS mortality (R-sq = 0.74). A combined index of HIV/AIDS, liver, homicide, and drug mortality rates showed R-squares above 0.5 with percent households with income less than $15K, Black and Hispanic percent of population, percent births to teenagers, births to single

The original version of this chapter was revised: Fig. 1.1 caption has been corrected now. The correction to this chapter is available at https://doi.org/10.1007/978-3-030-59624-8_6

mothers, and unemployment rate. Unemployment rate, median income, and percent low income households associated with the complex index with an R-square of 0.73.

In extremely high index CDs, the four related causes of death contributed 8 to 24% of total deaths. In most of these districts, diabetes also contributed over five percent. In extremely low index CDs, the four related mortality causes contributed one to three and a half percent of total deaths, and diabetes another one to three percent. The percent of the population over age 65 in the extremely high index CDs ranged from 5.1 to 11.5; in the extremely low index CDs, 8.3 to 18.8.

In spring 2020, coronavirus swept across New York City. Revisiting the CDs could convey insights about the local context of NYC's particularly fertile ground for COVID-19 and for the epidemic's possible aftermath.

1.2   Methods

All health data came from the 2017 *Vital Statistics Annual Summary*, the latest available from the NYC Department of Health: life expectancy; age-adjusted mortality rate; premature mortality rate (deaths under age 65); mortality rates of cancer, cerebrovascular disease, diabetes, drugs, flu/pneumonia, HIV, heart disease, homicide, liver disease, and chronic lower respiratory disease; infant mortality per 1000 live births; percent low-weight births, and percent births to teenagers—all by CD (NYC DOH 2019).

The 2017 CD poverty rate, unemployment rate, percent population over 65, percent adults with college or higher degree, percent housing with extreme over-crowding, serious housing violations per 1000 units, population density, median income, and percent households with rent stress (paying 35% income or more on rent) came from the Furman Center of New York University (Furman Center 2019). Fires by CD in 2017 came from the City Limits website (City Limits 2018). The online NYC Dept. of Planning Community Profiles provided 2010 poverty rate, unemployment rate, percent clean streets, percent households with rent stress, percent white, percent Latinx, percent black, population density, and percent adults with college or higher degree (NYC Dept. of Planning 2020).

Statgraphics Plus V5 analyzed the data: statistical description (average, median, standard deviation, minimum and maximum), t-test, Mann–Whitney test, bivariate regression, stepwise backwards multivariate regression, and dispersion (maximum/minimum).

Premature mortality rate associated with HIV, drug, and diabetes mortality rates and percent births to teenagers in multivariate regression. A combined index of these four rates was developed by normalizing each CD's rates with the medians of all CDs, weighting these normalized rates with the appropriate multiplicands in the equation produced by the multivariate regression, and adding together the four normalized, weighted rates for each CD.

1.3   Results

Mortality rates for heart disease and cancer dwarf all others (Table 1.1).

For these two mortality rates, the standard deviation is nowhere near the average or median. Homicide and HIV mortality rates present another picture: the standard deviation exceeds the median, an indication of large disparities. Although life expectancy has a very low standard deviation, the people in the longest living CD (CD 2Manhattan, includes Battery Park City) would live over a decade more than those in the shortest living CD (CD 16Brooklyn, Brownsville), another indication of large disparities.

Certain health outcomes show dispersions (maximum/minimum) over ten (Table 1.1). Diabetes, HIV, and homicide have minima of zero and theoretical dispersions of infinity, but their maxima—61 (diabetes), 16 (HIV), and 12.2 (homicide)—support a conclusion that these specific mortality rates show large disparities. Drug mortality rate, infant mortality rate, and percent births to teenagers have dispersions of 11.44, 11.17, and 74.00 respectively. The dispersion index of premature mortality is over twice that of overall mortality.

Premature mortality rate associates significantly in bivariate regression with many health outcomes (Table 1.2), four with R-squares over 0.5: percent births to

Table 1.1 Basic statistics of health outcomes: 59 community districts

Health outcome	Average	Median	SD	Min	Max	Dispersion
Cancer[a]	135.6	131.3	33.3	82.2	268.1	3.26
Cerebrovascular[a]	20.7	20.6	7.1	7.8	41.6	5.33
Diabetes[a]	20.6	18.4	12.1	0	61.7	Infinite[b]
Drugs[a]	16.9	14.8	9.3	3.9	44.6	11.44
Flu/pneumonia[a]	21.6	21.3	8.2	5.6	42.7	7.63
Heart[a]	189	175.5	70.7	72.2	492.6	6.82
HIV[a]	4.5	3.2	4	0	16	Infinite[b]
Homicide[a]	3.2	2.6	2.7	0	12.2	Infinite[b]
Infant mortality	4	4	1.6	0.6	6.7	11.17
Life expectancy	82	82	2.8	75.6	86.7	1.15
Liver[a]	6.3	5.6	3.1	1.6	14.2	8.88
Lower-respiratory[a]	19.4	18.8	7.4	5.5	41.9	7.62
Low-weight birth %	8.6	8.5	1.8	4.3	12.3	2.86
Mortality rate	520.1	514.4	121.9	311.7	809.4	2.60
% over 65	14.1	13.5	3.8	7.4	23.7	3.20
Premature mortality	172.4	157.9	69.6	63	379.1	6.02
% births to teenagers	3.2	2.9	2	0.1	7.4	74.00

[a]Crude rate per 100,000
[b]Max divided by zero (min) = infinite
Other rates are as conventionally measured
Premature dispersion/mortality dispersion = 2.32

Table 1.2 Associations of health outcomes with premature mortality rate

Health outcome	R-sq	P	
Percent teen births	0.6746	<0.0001	
Drug mortality	0.6690	<0.0001	
HIV mortality	0.6310	<0.0001	
Diabetes mortality	0.5674	<0.0001	
Homicide	0.4351	<0.0001	
Low-weight birth %	0.3760	<0.0001	
Infant mortality	0.3582	<0.0001	
Liver mortality	0.3027	<0.0001	
Cerebrovascular mortality	0.2285	0.0001	
Flu/pneumonia mortality	0.2052	0.0002	
Over age 65 %	0.2014	0.0002	Negative

Stepwise regression results
Premature mortality = 42.21+9.39 (teen birth%) + 2.07 (diabetes) + 2.76 (drugs) + 2.60 (HIV)
R-sq = 0.92

teenagers and mortality rates from diabetes, drugs, and HIV. Backwards stepwise regression shows that these four health outcomes together 'explain' 92% of the variability of premature mortality rate over the CDs. See the bottom of Table 1.2.

For the geography of ranges of premature mortality rate, see Fig. 1.1.

Figure 1.1 displays the wide difference in premature mortality rates in Manhattan and the Bronx. Nine of Manhattan's CDs have rates below 150/100,000, but none of the Bronx CDs have such low-ranking rates.

The difference between Brooklyn and Queens also stands out: Queens has only one CD out of fourteen with a rate above 200 whereas seven out of the eighteen CDs of Brooklyn have such high rates.

Seven NYC CDs have premature mortality rates below 100; nine have rates above 250. The group of seven low-rate CDs have an average median income of $115,721; the nine high-rate CDs have an average median income of $36,427. Other SE factors also parcel out starkly, including educational attainment, poverty rate, housing violations per 1000 units, percent white, percent black, percent Latinx, etc.

The socioeconomic and urban environmental factors also show varying dispersions (Table 1.3). With a range of 1–87.1%, black proportion of population shows the greatest dispersion; white proportion shows extremely high dispersion (range of 1.2–83.4%). Serious housing violations per 1000 units ranges from 3.2 to 125.4. Although percent adults with college or higher degrees had slightly higher dispersion in 2017 than in 2010, both poverty rate and unemployment rate rose steeply in dispersion between 2010 and 2017.

Backwards stepwise regression identifies college 2010, poverty rate 2017, housing violations per 1000 units, and percent units with extreme housing overcrowding as the independent SE variables that jointly associate with premature mortality rate pattern (Table 1.4).

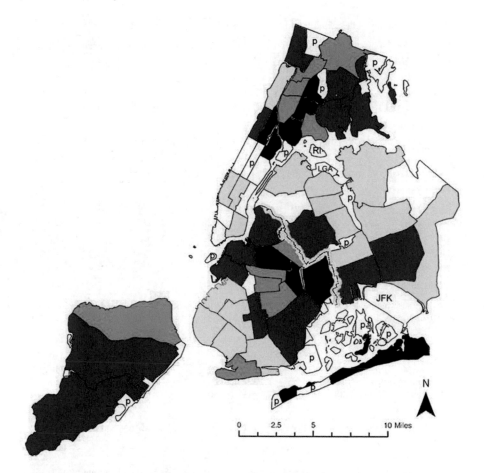

Fig. 1.1 Premature mortality rates of the Community Districts. Deaths per 10^5 for people under 65. Black = over 250, Blue = 200–250, Red = 150–200, Yellow = 100–150, White = under 100. P = park area, JFK = Kennedy Airport, LGA = LaGuardia Airport, Ri = Rikers Island

The stepwise regressions for members of the health outcome guild of premature mortality rate yields the following results (Table 1.4).

The variable 2017 unemployment appears as an independent variable in three of the four results; percent black 2010, in two; housing violations/1000 units in two; and one each for percent white 2010, percent Latinx 2010, median income, poverty rate 2017, and percent adults with college or higher degree 2010.

When the combined index of the four guild members is regressed against the SE factors, neither percent black 2010 nor percent Latinx 2010 reaches an R-sq of 0.4 (Table 1.5).

The single highest R-sq comes from housing violations per 1000 units, followed by unemployment 2017 and unemployment 2010. Poverty 2017 has an R-sq above 0.6. Backwards stepwise multivariate regression identifies percent adults with

Table 1.3 Statistical description and dispersions of socioeconomic factors

Factor	Average	SD	Minimum	Maximum	Median	Dispersion
Black % 2010	21.95	22.46	1	87.1	12.1	87.1
Clean streets %	94.00	3.07	85.1	99.1	94.3	1.16
College % 2010	35.98	20	10.3	82.3	31.1	7.99
College % 2017	36.43	19.5	9.7	83.7	32.2	8.63
Fires/100,000[a]	477.5	229.9	50.7	1733.9	455.3	34.20
Housing violations/1000 units	48.9	37.1	3.2	125.5	31.6	39.19
Latinx % 2010	29.8	20.5	6.4	71.3	22.9	11.14
Median household income ($)	64,492	30,261	20,640	147,640	57,680	7.15
Poverty rate 2010	19.9	6.8	7.2	35.6	20.2	4.94
Poverty rate 2017	19.1	10.2	6.1	44.2	16.2	7.25
Rent stress 2010	45.1	7.4	28.5	61	45	2.14
Rent stress 2017	29.9	6.3	16.8	45.6	30.5	2.71
Unemployment 2010	5.0	1.7	2.1	8.6	4.9	4.10
Unemployment 2017	6.9	3.7	2.1	20.7	5.8	9.86
White % 2010	32.1	24.5	1.2	83.4	25.7	69.5

[a]The highest incidence of fire is an outlier and very far above others. It also occurred in a wealthy area (midtown east). It is used here but may be a Fire Department typo

Table 1.4 Multivariate equations of SE factors for premature mortality and its guild

Premature mortality rate = 184.53 − 1.22 (% college 2010) − 12.15 (EOC %) + 0.565 (housing violations/1000 units) + 3.086 (poverty 2017)

R-sq = 0.80

Diabetes mortality rate = 15.42 + 0.47 (% black 2010) − 0.00018 (median income) + 1.96 (%white 2010)

R-sq = 0.69

Drug mortality rate = 2.45 + 0.443 (poverty rate 2017) + 0.865 (unemployment 2017)

R-sq = 0.60

HIV mortality rate = −0.65 + 0.04 (% black 2010) + 0.0045 (housing violations/1000 units) + 0.29 (unemployment 2017)

R-sq = 0.63

Teen birth % = 2.13 − 0.035 (college 2010) + 0.019 (housing violations/1000 units) + 0.022 (% Latinx 2010) + 0.104 (unemployment)

R-sq = 0.85

Dependent variable	Independent variables
Diabetes	Percent black 2010, median income, percent white 2010
Drugs	Poverty 2017, unemployment 2017
HIV	Percent black 2010, housing violations/1000 units, unemployment 2017
Teen birth %	Percent college 2010, housing violations/1000 units, percent Latinx 2010 unemployment 2017

College 2010: percent adults with college or higher degree in 2010
EOC extreme housing overcrowding

Table 1.5 Associations of the combined guild index with SE factors

Factor	R-sq	P	pos/neg
Black 2010	0.3701	<0.0001	pos
Clean streets	0.2818	<0.0001	neg
College 2010	0.4823	<0.0001	neg
College 2017	0.4716	<0.0001	neg
Fires/100,000	0.0555	0.0398	pos
Housing violations/1000 units	0.7147	<0.0001	pos
Latinx 2010	0.3711	<0.0001	pos
Median income	0.5336	<0.0001	neg
Poverty 2010	0.5298	<0.0001	pos
Poverty 2017	0.6111	<0.0001	pos
Rent stress 2010	0.2145	0.0001	pos
Rent stress 2017	0.2239	0.0001	pos
Unemployment 2010	0.6438	<0.0001	pos
Unemployment 2017	0.6758	<0.0001	pos

Table 1.6 Comparisons of SE factors of CDs above and below premature mortality median

SE Factor	SE medians compared		P
	Above premature median	Below premature median	
% black	29.3	4.3	1.00E-07
% Latinx	37.1	15.8	0.0129
% white	14.4	46.1	5.20E-05
% clean streets	93	95.4	0.0134
% college 2010	24.1	39.6	0.0001
% college 2017	27.3	39.2	0.0002
Fires/10E5	524.4	351.2	0.0002
Housing violations[a]	75	20.7	1.80E-07
Median income $	50,290	70,620	2.00E-05
Poverty 2010	21.2	17.7	0.0017
Poverty 2017	24.4	13.2	0.0002
Rent stress	32.1	29.7	0.037
Unemployment 2010	6.2	3.75	8.80E-07
Unemployment 2017	8.2	4.8	8.20E-07

[a]Per 1000 units

college education 2010, housing violations/1000 units, and unemployment 2017 as the associated independent SE factors for the combined index (R-sq = 0.83).

The median premature mortality rate is 157.9. Division of CDs into those above and below that median can yield informative statistics (Table 1.6).

Percent black 2010, percent adults with college or higher degree 2010, housing violations per 1000 units, poverty rate 2017, unemployment 2010, unemployment 2017, and percent white 2010 showed median differences between the sets of CDs with extremely low P's of 10^{-5}–10^{-7}.

Some less starkly different SE factors presented marked contrasts between the sets of CDs. Household median income had medians of $50,290 and $70,620 for CDs above and below premature mortality median, respectively. The median for percent Latinx 2010 is 37.1 for the higher CDs and 15.8 for the lower. The higher CDs had on median 524.4 fires per 100,000, and the lower, 351.2. The 2000 mortality rates of HIV, homicide, drugs, liver disease, and diabetes can be compared with those from 2017 (Table 1.7).

Mortality rates of both HIV/AIDS and homicide declined greatly between 2000 and 2017, but drug-related mortality rates slightly increased and those from liver disease and diabetes showed a little change.

1.4 Discussion and Conclusion

HIV/AIDS in 2000 contributed over 10% of total deaths in three Bronx CDs; its unweighted mean mortality incidence was 26.4 per 100,000 over the 59 CDs (Wallace et al. 2016). The guild of mortality causes (HIV/AIDS, homicide, liver, and drugs) together contributed over 20% of total deaths in two Bronx CDs. In 2000, HIV and its mortality guild posed serious public health threats to poor neighborhoods of color and played major roles in premature death.

Although HIV-related deaths show vast improvement (unweighted mean 2017 incidence/100,000 of 4.5), it shows high dispersion: minimum of zero; maximum of 16. Diabetes mortality, drug mortality, homicide, infant mortality, and percent births to teenagers also showed dispersion above 10, with percent births to teenagers topping the list at a dispersion of 74.

The CD pattern of premature mortality is almost entirely associated (R-sq = 0.92) with the joint pattern of percent births to teenagers and mortality rates of diabetes, drugs, and HIV. Eighty percent of the CD pattern of premature mortality is associated with four SE factors: poverty rate 2017, college or higher degrees per hundred adults in 2010, housing violations per 1000 units, and percent units with extreme overcrowding. Premature mortality and its guild of health outcomes associate with poverty rate, indicators of segregation, educational attainment, dangerously substandard housing, and unemployment.

The only non-mortality rate among the guild for premature mortality is percent births to teenagers. Births to teenagers may indicate premature aging in communities of color. Black women age faster than white women, as indicated by declining weights of babies born to black women in higher age brackets (Geronimus 1996). The historic dynamic of high proportions of black teenagers having babies no longer holds true in New York, a change reflected in the reduction in R-square in the relationship between percent black and percent low-weight births from 0.52 in 2010 to 0.25 in 2017. Percent Latinx has an R-sq of 0.53 in regression with percent births to teenagers in 2017.

In New York City, similar dynamics to the historic ones among black women may dominate presently among the Latin American women. Eleven CDs had premature

Table 1.7 Comparison of mortalities 2000 and 2017

Mortality rate	2000[a]					2017				
	Mean	Median	SD	Minimum	Maximum	Mean	Median	SD	Minimum	Maximum
HIV/AIDS	26.4	17.2	23.2	0.9	99.2	4.5	3.2	4	0	16
Drugs	11.1	9.6	7.3	2.1	35	16.9	14.8	9.3	3.9	44.6
Liver	7.1	6.4	4.3	1.7	23.3	6.3	5.6	3.1	1.6	14.2
Homicide	9.1	8.1	6.5	0.8	24.8	3.2	2.6	2.7	0	12.2
Diabetes	23.4	21.2	11.8	2.9	63.4	20.6	18.4	12.2	0	61.7

[a]From Wallace et al. (2016)

mortality rates of 240 or above. In seven of these, Latinx percent of population was 46 or above (six with over 60%, Bx CDs 1–6). Of the eleven, only one had a percent teen births of less than 4 (Central Harlem). We can conclude that the CDs with high premature mortality rates also have high rates of teen births and that the great majority of these CDs are predominantly Latinx. Teenaged girls in Latinx communities may be having babies while their own mothers are still functionally young enough to offer effective help, a relationship similar to that in black communities in the past (Stack 1974; Geronimus et al. 1999).

When we contrasted the seven CDs with very low premature mortality rates with the nine with very high rates, the class and race/ethnicity structure of basic health stood out clearly. The SE factors of the two groups of CDs differed greatly: median income, educational attainment, poverty rate, housing violations per 1000 units, percent white, percent black, percent Latinx, etc.

SE associations of rates of births to teenagers hint that these rates index particular social and economic pressures on families and whole populations, pressures from unemployment, poverty, housing deficits, and discrimination. Rates of births to teenagers continue to index forces that either foster 'weathering' or buffer against it. Rates of births to teenagers have declined steadily over the decade in New York City as a whole (NYC DOH 2019). Geographic areas that rank high for this health outcome suffer from economic, social, and discriminatory pressures. This health outcome and percent low-weight births together indicate accelerated aging ('weathering'). Percent births to teenagers has the highest R-sq of any of the health outcomes in the CD database in bivariate regression with percent low-weight births. Age is the greatest risk factor for mortality from COVID-19. 'Weathering' determines physiological age.

New York City's Department of Health (DoH) defines premature death as one taking place below age 65. For a long time now, the National Center for Health Statistics has used age 75 as the benchmark and totes up years of life lost below age 75. In failing to change its criterion, New York masks the true geographic and population extent and intensity of the problem of premature death. All CDs have life expectancies above 75. The lowest is 75.9. There is no scientific reason for this clinging to the old criterion, but changing to the one in use at the Federal level would raise premature mortality rates immensely. In using the City's criterion in this paper, we, in a sense, collaborate in the smokescreen. The City makes only certain data available at the CD level, one of them being its peculiarly calculated premature mortality rate. We acknowledge and apologize for this shortfall which is both scientific and ethical.

The particular socioeconomic factors that associate with premature mortality rate and each member of its guild hint at the constellation of stressors producing early death: unemployment, poverty, bad housing, a low proportion of adults in the community with higher educational attainment, and segregation. In 2000, the three independent SE variables that associated in multivariate regression with the combined index of the HIV guild were unemployment, median income, and percent of households with income of $15,000 or less (Wallace et al. 2016). Other factors with high R-squares in regression with the combined index were percent black and percent white. In brief, the SE associations in 2000 with the HIV guild closely resembled those in 2017 with premature mortality rate and its guild of health outcomes.

In this study, serious housing violations per 1000 units shows up as an independent variable in the multivariate regressions of SE factors with percent births to teenagers, with HIV mortality rate and with premature mortality rate. Epidemiologists have long regarded housing access, security, and quality as strong influences on public health. Recent studies concentrated on serious violations as indicators of household allergens in asthma etiology (Rosenfeld et al. 2010; Beck et al. 2014). O'Campo et al. (2000) examined injury rates among children as outcomes of housing code violations. Housing inspections may serve as potential interventions for specific public health problems such as lead poisoning (Korfmacher and Holt 2018). However, Shaw (2004) and Krieger and Higgins (2002) viewed the relationship between housing access and quality on one hand and public health on the other much more broadly and concluded that housing forms a key determinant of public health. Krieger and Higgins listed details of particular public health problems ranging from childhood asthma due to pest allergens to elevated mortality rates in the elderly living in cold units. They came to the conclusion that public health scientists and practitioners should advocate for adequate housing supply and quality.

Very recent studies on unemployment documented a wide range of health outcomes: increased all-cause and specific-cause mortality rates (Nie et al. 2020), mental health disorders and risk behaviors in young adults (Vancea and Utzet 2017), development of type 2 diabetes (Pinchevsky et al. 2020), failing in follow-up for antiretroviral treatment for HIV (Frijters et al. 2020), and increased child abuse by unemployed parents (van Berkel et al. 2020). Percent adults unemployed in 2000 had the highest R-sq in association with the HIV-related mortality guild (Wallace et al. 2016). Three independent variables associated with the pattern of the combined index of the HIV-related guild: unemployment rate, median income, and proportion of households with income of $15K or below. The 2017 relationship between community premature mortality rate and unemployment rate is mediated by risk behaviors that are reinforced or mitigated by other SE factors such as community median income, community poverty rate, and community educational attainment.

Poverty rate 2017 also influenced premature mortality rate pattern. Recent studies on links between health and poverty reveal a broad range of mechanisms. Food insecurity leads to overcompensation in eating and higher rates of obesity in parents and children (Brown et al. 2020). Number of types of insecurity influences probability and severity of depression (Wallace et al. 2003). Failure to take medicine (Hensley et al. 2018) and indulgence in various risk behaviors (Pantell et al. 2019) have been associated with poverty on individual and community levels. Evictions have strong effects on children's mental health (Hazecamp et al. 2020). Housing overcrowding raises probabilities of within household spread of disease (example: Drucker et al. 1994). Chronic stress erodes sleep quality and quantity which, in turn, changes metabolism toward visceral fat deposition (Spiegel et al. 2009); chronic stress dysregulation of the HPA axis brings on metabolic syndrome and visceral fat deposition and the cascade of chronic conditions therefrom such as coronary heart disease, diabetes, and cerebrovascular disease (Arnetz and Ekman 2006).

The final common influence is college or higher degrees per hundred adults. Differing rates of college education among the states associate with differing mortality rates, life expectancies, and many cause-specific mortality rates (Wallace and Wallace 2018, 2019). Coronary heart mortality rates in 2014 associated negatively over the 50 states with percent adults with college or higher degrees in 2000 for age ranges 45–54, 55–64, and 65–74 with the following respective R-squares: 0.55, 0.50, and 0.43. Diabetes mortality rates over the 50 states associated negatively with percent adults with college or higher degrees in age ranges 45–54, 55–64 and 65–74 with the following R-squares: 0.49, 0.42, and 0.42. Similarly, the following health outcomes associated negatively with percent adults with college or higher degrees with R-squares above 0.4: infant mortality rate; mortality rates of children in age ranges 1–4, 5–9, and 10–14; percent adults not eating fruit daily; smoking prevalence; incidence of teen births; and vehicle fatality incidence (Wallace and Wallace 2018, 2019).

Although ability to understand and apply information from health authorities likely plays a role in the health uplift from higher education for individuals and communities, the literature hints at deeper mechanisms. Highly educated Palestinians show more resilience in the face of chronic crisis (Kteily-Hawa et al. 2020). Educational attainment has long been acknowledged as a buffer against organic damage from Alzheimer's disease, often called the cognitive reserve (Yasuno et al. 2020). Educational attainment, independent of cognitive function, protects against coronary heart disease and stroke (Gill et al. 2019).

Higher education may actually restructure the brain for adaptability of function as the phenomenon of 'cognitive reserve' hints. It also confers a deeper layer of cultural resource, increased critical thinking, access to wider sources of information, habits of gleaning wider sources of information, and a greater sense of mastery over circumstances. Higher education may set individuals and their families on a trajectory of wiser habits that shaves the community-level incidence of premature mortality. Communities with critical masses of highly educated residents likely enjoy a more rapid and profound flow of vital information and creative adaptability than those without this human resource.

The late nineteenth to early twentieth century Great Reform led to immense improvements in public health: decline in tuberculosis incidence and mortality rate (Fairchild and Oppenheimer 1998), decline in child mortality rates (Field and Behrman 2003), decline in cholera incidence (Pizzi 2002), and many other improvements. Life expectancy showed steady increases (Statistica 2020). With the policies and practices from Nixon on, such markers of countervailing forces (Galbraith 2010) as labor union participation and civil rights enforcement declined (Henderson 2009). Even many professionals, proletarianized by the 'gig economy', have no job security, benefits such as health insurance and pensions, or control over working conditions.

Griscom (1844) and Chadwick (1842) organized the foundations of the Great Reform in the United States and in England. Griscom described his early interactions with churches, charities, and other institutions with activities in the laboring-class neighborhoods. A hundred years later, Dubos and Dubos (1952) outlined programs needed to minimize TB incidence and prevalence: education, community organizing, labor organizing, and general strengthening of what Galbraith (2010) termed 'countervailing forces', the entities that keep quasi-monopolies from absolute control of society and the economy. The Great Reform and its successor, the human rights campaigns in post-war times, focused on empowering of the powerless through education and organizing.

Directly observed therapy with antibiotics controls tuberculosis outbreaks in the United States (Chaulk et al. 1995), but the conditions that Dubos and Dubos (1952) described as attendant on the 'Captain of All the Men of Death' creep into the context of premature mortality step by step. Loss of low-cost housing, beginning in the Richard Nixon Administration (Sisson et al. 2019), over many municipalities resulted in a housing famine exacerbated by unchecked gentrification (Lartey 2018). Overcrowding and homelessness again arose as a consequence (National Institute of Medicine 1988). Loss of good unionized factory jobs left workers without college degrees either underemployed or unemployed (Bluestone 1983). Nixon's Southern Strategy shifted jobs to southern and western states with right-to-work laws, but the shift was only temporary as globalization gutted US manufacturing, eroding those shifted jobs (Wallace and Wallace 2019).

Medicine has, in this post-Great Reform era, been used to keep many socially rooted health problems under fragile control. Everything from antibiotics to statins keep Americans (and British) from racking up impressive mortality rates from diseases and chronic conditions that arise in the context of bad working and living conditions and from the uncertainty and fear due to feudal power relations. This papering over of cracks in public health and safety began losing its mythology shortly after the turn from twentieth to twenty-first century when reports on lower life expectancy among white people without college education began appearing in the literature (Olshansky et al. 2012). Indeed, medicine contributed to the growth of the public health crisis with such new ills as an epidemic of prescription opioid addiction (CDC drug overdose on website) and the rise of antibiotic resistance from both medical and agricultural misuse (WHO 2018).

Our data here show that in present New York City, the old enemies of public health and long life exert their force with renewed vigor: poverty, unemployment, lack of education, substandard housing, and segregation. This intertwined context of SE factors and public health forms the stage onto which the COVID-19 pandemic entered. The public policies at federal, state, and municipal levels that set this stage gathered steam from Nixon through Barack Obama and Donald Trump at the federal level; from Nelson Rockefeller through Andrew Cuomo at the state level; and from Lindsay through Bill de Blasio at the municipal level—with the cooperation of the legislatures.

Many public health scientists and practitioners see the COVID-19 crisis as a turning point in American society. The vision of rolling back the power grab of the '1%' arises in many an OpEd, a restructuring ala Isaiah that exalts every valley and makes low every mountain and hill. Andrew Cuomo, governor of New York State, in a coronavirus update press conference, envisioned a vast restructuring of society with greater equality in economics and politics and less favoring of the wealthy (Guardian 2020). His 2020/2021 budget, however, reflects continued dominance of the wealthy and the privileged in its failure to spread the pain of a massive shortfall over the social classes (Gothamist 2020). Wealthy individuals continue in their current tax rates with all their deductions, and large wealthy corporations continue in their tax abatements. Medicaid is slashed as are funds for municipalities. Education is static. Bail and discovery reforms are rolled back and will result in more people held pre-trial in jail and convicted to prison during trial under lack of access to evidence held by prosecutors, and the promised parole reform is scrapped. The NYS budget director gets the unprecedented power to cut funding monthly in any year that the tax receipts fall below 99% of those projected. This budget reinforces the feudal hierarchy. And it is real feudalism (Wallace and Wallace 2019).

The Great Reform did not gain traction until after the Triangle Shirtwaist Fire. It did not gain ascendance until the New Deal programs to buffer effects of the Great Depression. Bringing another Reform to reality will likewise be a lengthy, laborious, dangerous, often-tedious, and demoralizingly frustrating piece of work. Without this new Great Reform, the old enemies of public health and well-being will leave the people of NYC and of America sitting ducks for pandemics and premature mortality between pandemics: poverty, unemployment, low educational attainment, substandard and insufficient housing, and the war on countervailing forces such as labor unions, public health and environmental scientists, human rights groups, and voters.

The largest differences in SE factors between CDs above and below the median for premature mortality were % black, % white, housing violations per 1000 units, median income, and unemployment rates. The combination of segregation of black New Yorkers and discrimination against black and Latino New Yorkers yielded the substandard living conditions and narrowed opportunities for employment and good wages demonstrated in Table 1.6. Black and Latino New Yorkers suffer from greatly thwarted social mobility, a major component of hope as opposed to the despair now being discussed in public health. Social mobility via education and training declined greatly since 1980 (Davis and Mazumder 2017).

Social mobility mixes classes within families and communities. It is a goal of many labor unions, even to the point of offering college scholarships to members' children. Social mobility is so powerful that black women born in poverty but gaining middle class status as adults are not plagued with the low-weight births that mark 'weathering' (Love et al. 2010). Without a return to at least 1960s–1970s rates of social mobility for everyone, there will be no return to steady improvements in public health and life expectancy. Without increasing social mobility, the high percent of the American population vulnerable to infection and mortality from new animal-to-human diseases will ensure that these diseases roll through the populace.

When communities offer good housing, good schools, good essential services, and residential stability, the sons and daughters who achieve higher educational attainment and the employment associated with it will take up their adult homes in the old neighborhoods. Their friends and families are there. Communities with inadequate and substandard housing, underfunded schools, throttled essential services, and high residential in-and-out migrations don't invite young high achievers to remain as adults. These communities cannot enjoy the resources offered by a critical mass of residents with college or higher degrees. Hierarchical and feudal public policies and FIRE (Finance, Insurance and Real Estate) industry practices keep them caught in the vicious circle of poverty, unemployment, and discrimination. Governmental officials and FIRE industry executives conspire to retain the rigid hierarchy with laws, enforcement, and regulations tilted heavily toward the wealthy, as Andrew Cuomo's budget for 2020/2021 demonstrates.

New York City is an extreme example of fertile ground for contagion, but not a complete outlier. It has long served as the summit of the national and global metropolitan regional network and blows back out to the other metropolitan regions the contagious diseases and behaviors that enter it (Wallace et al. 1997). As Chap. 3 describes, the local vulnerabilities couple with the conduits of contagion (the contacts between counties in a metropolitan region and between metropolitan regions) to produce larger scale patterns of morbidity and mortality. Every American metropolitan region has fertile ground for contagion. That is why the Black Lives Matter movement has gained traction. The police violence and corruption required to retain the hierarchy contribute to the premature mortality rates in black communities.

References

Arnetz B, Ekman R (2006) Stress in health and disease. Wiley, Wienheim

Beck A, Huang B, Chundur R, Kahn R (2014) Housing code violation density associated with emergency department and hospital use by children with asthma. Health Aff 33(11):1993–2002

Bluestone B (1983) Deindustrialization and unemployment in America. Rev Black Polit Econ 12(3):27–42. https://doi.org/10.1007/BF02873944

Brown C, Skelton J, Palakshappa D, Pratt K (2020) High prevalence of food insecurity in participants attending weight management and bariatric surgery programs. Obes Surg 30(9):3634–3637 https://doi.org/10.1007/s11695-020-04645-7. E-pub ahead of print

Carey B, Glanz J (2020) Travel from city seeded spread nationwide. NY Times. May 8: P 1, 9

CDC (2019) https://www.cdc.gov/coronavirus/2019-ncov/cases-updates/cases-in-us.html. Accessed 10 May 2020

CDC (2020) Drug overdose on website. Opioid overdose. Understanding the epidemic. https://www.cdc.gov/drugoverdose/epidemic/index.html

Chadwick E (1842) Report from the Poor Law Commission on an Inquiry into the Sanitary Condition of the Laboring Population of Great Britain. https://ia902707.usarchive00chadwick/reportonsanitary/chadwick.pdf

Chaulk C, Moore-Rice K, Rizzo R, Chaisson R (1995) Eleven years of community-based directly observed therapy for tuberculosis. JAMA 274(12):945–951

City Limits (2018) https://citylimits.org/2018/02/13/urbanerd-where-nyc-fires-were-in-2017

Davis J, Mazumder B (2017) The decline in intergenerational mobility after 1980. Opportunity and Inclusive Growth Institute. Federal Reserve Bank of Minneapolis. Working Paper 17–21. July 2017. https://www.minneapolisfed.org/institute/working-papers/17-21.pdf

Drucker E, Alcabes P, Bosworth W, Sckell B (1994) Childhood tuberculosis in the Bronx, New York. Lancet 343(8911):1482–1485

Dubos R, Dubos J (1952) The White Plague: tuberculosis, man and society. Little, Brown, and Company, Boston

Fairchild A, Oppenheimer G (1998) Public health nihilism vs. pragmatism: history, politics, and the control of tuberculosis. Am J Public Health 88(7):1105–1117

Field M, Behrman R (eds) (2003) Chapter 2. Patterns of childhood death in America. In: When children die: palliative and end-of-life care for children and their families. Institute of Medicine. National Academies Press (US), Washington DC

Frijters E, Mermans L, Wensing A, Deville W, Tempelman H, DeWit J (2020) Risk factors for loss to follow-up from antiretroviral therapy programmes in low- and middle-income countries: a systematic review and meta-analysis. AIDS 34(9):1261–1288. https://doi.org/10.1097/QAD. 0000000000002523 (Epub ahead of print)

Furman Center (2019) https://furmancenter.org/neighborhoods. Accessed 8 April 2020

Galbraith J (2010) Reprinted 2010. Introduction to American capitalism. Literary Classics of the United States, New York, pp 5–6

Geronimus A (1996) Black/white differences in the relationship of maternal age to birthweight: a population-based test of the weathering hypothesis. Soc Sci Med 42:589–597

Geronimus A, Bound J, Waidmann T (1999) Health inequality and population variation in fertility-timing. Soc Sci Med 49(12):1623–1636

Gill D, Efstathiadou A, Cawood K, Tzoulaki I, Dehghan A (2019) Education protects against coronary heart disease and stroke independently of cognitive function: evidence from Mendelian randomization. Int J Epidemiol 48(5):1468–1477

Gothamist (2020) https://Gothamist.com/news/Cuomo-state-budget-robust-progressives-call-it-republican-austerity-warfare

Griscom J (1844) The sanitary condition of the laboring population of New York. Arno Press, New York. Reprinted 1970

Guardian (2020) https://www.theguardian.com/world/live/2020/Apr/27/coronavirus-us-live-cases-america-trump-cuomo

Hazecamp C, Yousuf S, Day K, Daly M, Sheehan K (2020) Eviction and pediatric health outcomes in Chicago. J Community Health 45(5):891–899. https://doi.org/10.1007/s10900-020-00806-y

Henderson W (2009) A strong labor movement is critical for civil rights. Testimony of Wade Henderson before the senate committee on health, education, labor, and science. March 10, 2009. https://civilrights.org/resource/a-strong-labor-movement-is-critical-for-civil-rights-testimony-of-wade-henderson

Hensley C, Heaton P, Kahn R. Luder H, Frede S, Beck A (2018) Poverty, transportation, and medication nonadherence. Pediatrics 141(4):e20173402. https://doi.org/10.1542/peds.2017-3402. Epub March 16

Korfmacher K, Holt K (2018) The potential for proactive housing inspections to inform public health interventions. Public Health Manag Pract 24(5):444–447

Krieger J, Higgins D (2002) Housing and health: time again for public health action. Am J Public Health 92(5):758–768

Kteily-Hawa R, Khalifa D, Abuelaish I (2020) Resilience among a large sample of adult Palestinians in the Gaza Strip: examining contextual sociodemographic factors and emotional response through a social-ecological lens. Public Health 182:139–142

Lartey J (2018) Nowhere for people to go: who will survive the gentrification of Atlanta? The Guardian. https://www.theguardian.com/cities/2018/oct/23/nowhere-for-people-to-go

Love C, David R, Rankin K, Collins J (2010) Exploring weathering: effects of lifelong economic environment and maternal age on low birth weight, small for gestational age, and preterm birth in African-American and white women. Am J Epidemiol 172(2):127–134. https://doi.org/1093/aje/ewq/109-1-8

National Institute of Medicine (1988). Chapter 2. Dynamics of homelessness. In: Homelessness, health, and human needs. National Institute of Medicine (US) Committee on Health Care for Homeless People. National Academy Press (US), Washington DC

Nie J, Wang J, Aune D, Huang W, Xiao D, Wang Y, Chen X (2020) Association between employment status and risk of all-cause and cause-specific mortality: a population-based prospective cohort study. J Epidemiol Community Health 74(5):428–436

NYC Department of Health (2019) Summary of vital statistics: 2017. https://www1.nyc.gov/assets/doh/downloads/pdf/vs/2017sum.pdf

NYC Dept. of Planning (2020) Accessed 2020. https://communityprofiles.planning.nyc.gov

NYS (2020) https://COVID19tracker.health.ny.gov/views/ NYS-COVID19-tracker/NYSDOHCOVID19Tracker-Map. Accessed 10 May 2020

O'Campo P, Rao R, Gielen A, Royalty W, Wilson M (2000) Injury producing events among children in low-income communities: the role of community characteristics. J Urban Health 77(1):34–49

Olshansky S, Antonucci T, Berkman L et al. (2012) Differences in life expectancy due to race and educational differences are widening, and may not catch up. Health Aff (Millwood) 31(8):1803–1813

Pantell M, Prather A, Downing J, Gordon N, Adler N (2019) Association between social and behavioral risk factors with earlier onset of adult hypertension and diabetes. JAMA Netw Open 2(5):e193933. https://doi.org/10.1001/jamanetwopen.2019.3933

Pinchevsky Y, Butkow N, Raal F, Chirua Y, Rothberg A (2020) Demographic and clinical factors associated with development of Type 2 diabetes: a review of the literature. Int J Gen Med 13:121–129

Pizzi R (2002) Apostles of cleanliness. MDD 5(5):51–55

Rosenfeld L, Rudel R, Chew G, Emmons K, Acevedo-Garcia D (2010) Are neighborhood-level characteristics associated with indoor allergens in the household? J Asthma 47(1):66–75

Shaw M (2004) Housing and public health. Annu Rev Public Health 25:397–418

Sisson P, Andrews J, Bazeley A (2019) The affordable housing crisis, explained. https://www.curbed.com/2019/5/15/18617763/affordable-housing-policy-rent-real-estate-apartments

Spiegel K, Leproult R, van Cauter E (2009) Effects of poor and short sleep on glucose metabolism and obesity risk. Nat Rev Endocrinol 5:253–261

Stack C (1974) All our kin: strategies for survival in a black community. Harper and Row, New York

Statistica (2020) Graph of life expectancy 1860–2020 from birth in the United States. https://www.statista.com/statistics/1040079/life-expectancy-united-states-all-time

van Berkel S, Prevo M, Linting M, Dannebakker F, Alink L (2020) Prevalence of child maltreatment in the Netherlands: an update and cross-time comparison. Child Abuse Negl 103:104439. https://doi.org/10.1016/jchabu.2020.104439. Epub ahead of print

Vancea M, Utzet M (2017) How unemployment and precarious employment affect the health of young people: a scoping study on social determinants. Scand J Public Health 45(1):73–84

Wallace D, Galea S, Ahern J, Wallace R (2016) Chapter 12. Death at an early age: AIDS and related mortality in New York City. In: Gene Expression and its discontents: the social production of chronic disease, 2nd edn. Springer, Cham

Wallace D, Wallace R (1998) A plague on your houses: How New York City was burned down and national public health crumbled. Verso Books, New York and London

Wallace D, Wallace R (2018) Right-to-work laws and the crumbling of American Public Health. Springer, Cham

Wallace D, Wallace R (2019) Politics, hierarchy, and public health: voting patterns in the 2016 US presidential election. Routledge International Studies in Health Economics, New York

Wallace D, Wallace R, Rauh V (2003) Community stress, demoralization, and body mass index: evidence for social signal transduction. Soc Sci Med 56:2467–2478

Wallace R, Huang Y, Gould P, Wallace D (1997) The hierarchical diffusion of AIDS and violent crime among US metropolitan regions: inner city decay, stochastic resonance, and reversal of mortality transition. Soc Sci Med 44:935–947

WHO (2018) Antibiotic resistance. https://www.who.int/news-room/fact-sheets/detail/antibiotic-resistance

Yasuno F, Minami H, Hattori H (2020) Interaction effect of Alzheimer's disease pathology and education, occupation, and socioeconomic status as a proxy for cognitive reserve on cognitive performance: in vivo positron emission tomography study. Psychogeriatrics. https://doi.org/10.1111/psyg.12552

Chapter 2
NYC COVID Markers at the ZIP Code Level

2.1 Introduction

The previous chapter analyzed New York City (NYC) geographic patterns of premature mortality at the Community District (CD) level. Most CDs encompass more than one ZIP Code. New York City has 59 CDs whereas Queens alone has 56 ZIP Code areas with more than 10,000 residents. The US Postal Service invented ZIP Codes decades ago to organize mail delivery efficiently. How far a mail carrier can walk and how much a mail carrier can carry or trundle go into determining the size and shape of ZIP Codes. Population density, housing crowding, and socioeconomic factors that influence volume of mail per unit population also contributed to decisions about size and shape of ZIP Code areas, factors summarized on the US Postal Service website. We know that a NYC ZIP Code represents an area that a mobile person can walk across and that likely shows some uniformity in socioeconomic characteristics. We also know that few NYC Zip Codes cross county boundaries and, thus, are arrayed in blocks of similarly administered larger zones, the boroughs. For our purposes, NYC ZIP Code areas present acceptable analytical geographic units for understanding the geography of COVID within boroughs.

Although the media tried to pinpoint a lawyer who lived in Westchester County and worked in Manhattan as the index COVID case for the New York City epidemic, the massive transportation system leaves the whole metropolitan region open to multiple 'patients zero' because the local, regional, national, and international linkages present numerous points of entry. Three clusters in Westchester County, the Bronx, and Manhattan may have gotten their starts with this lawyer, but unrelated clusters popped up in Queens, the Bronx, Brooklyn, and eventually Staten Island.

The original version of this chapter was revised: Figs. 2.1 and 2.2 captions have been corrected now. The correction to this chapter is available at https://doi.org/10.1007/978-3-030-59624-8_6

Manhattan and the Bronx have similarly sized populations (1.4–1.6 million), as do Brooklyn and Queens (2.2–2.5 million). Manhattan and the Bronx historically showed close epidemiological connections. The tuberculosis (TB) epidemic of 1979–1993 jumped from Harlem to the central South Bronx which became the epicenter for spread through the Bronx (Wallace 1994). However, Brooklyn and Queens, although geographically abutting, do not share these connections. Although the poverty belt of Brooklyn lies along the border with Queens, the poverty zones of Queens lie far from that border. Furthermore, travel between the two large boroughs by public transportation poses difficulties. One little NYC subway line (the G line) with a bad schedule and a couple of bus lines cross that border (MTA website). We decided to compare Manhattan with the Bronx and Brooklyn with Queens for COVID indicators (percent positive swab tests, cases per unit population, and mortalities per unit population) and for SE factors of potential influence. We also decided to select a few health outcomes for possible association with COVID indicators: premature mortality rate, and mortality rates of diabetes, heart disease, cancer, and flu/pneumonia.

COVID crested in New York City in April 2020 (NYC DoH 2020). By the end of May, the boroughs had accumulated crest and post-crest swab test results, cases, and deaths. The small differences in timing of crest were overcome by the end of May. Many of the characteristics of the pandemic in the context of New York had become apparent also. Nursing homes generated large numbers of serious cases and deaths among the elderly residents and many cases among the care-taking staff who were often not supplied with proper personal protective gear (Nursinghome411 2020). Intense early clustering in ethnic neighborhoods in Queens hinted at problems in health education among immigrants and within neighborhoods generally out-of-the-mainstream-loop. This accelerated start in Queens led to continuously higher boroughwide case numbers and rates in Queens than in Brooklyn, although Brooklyn has slightly more people. Case and mortality rates were higher among men than women and among those over age 60 than among younger age ranges (NYC DoH 2020). These age and gender patterns were also seen nationally (CDC 2020). Looking at small-area COVID indicators and the associated SE factors and health outcomes could illuminate the context of the case and death rate patterns.

2.2 Methods

Cumulative ZIP code area data on percent positive swab tests, cases per 100,000, and deaths per 100,000 came from the New York City Dept of Health website (DoH 2020) which is updated daily and which we accessed May 31, 2020. All ZIP code area SE data came from Infoshare NYC which had gathered them from the 2013–2017 American Community Survey. The Infoshare 2011–2013 mortality data came from records of the NYC Department of Health (DoH). Later DoH data were not available.

We calculated 'percent college or higher degree' as number of degrees per hundred adults 25 years old or older. Thus, if one person had a BA, an MA, and a Ph.D, that counted as three degrees. This is different from the US Census method which would count it as a single person with college or higher degree.

All population data come with some uncertainties. Small area data are especially sensitive to mistakes, ephemeral blips, and transitions. The COVID data in particular deserve a jaundiced view because they rely on testing. The early testing phases were non-uniform, having been initiated in panic, in ignorance of the virus's behavior, and with confounding restrictions that limited testing to the very ill. The later phases may have been applied in a biased manner to safeguard valued populations rather than everyone. We know that the data probably undercount the presence of infected individuals, cases, and deaths in traditionally underserved communities. However, they are all that exist and may retain enough information to indicate patterns.

We are aware that the 2011–2013 DoH death data may predate the current health profiles of the ZIP Code areas a bit too much. Small areas may experience rapid changes. Again, they are all that are publicly available now and may still present information of value.

In looking for a better index of 'weathering', we decided that premature mortality rate was too coarse because it includes all age ranges from birth to age 65. We found that the age range with two characteristics was a better index: (1) a rise in number of deaths above any in younger age ranges and (2) rises in numbers of deaths with each older age range above it. This age range marks the approximate beginning of 'weathering', aka physiological premature aging (Geronimus 1996). Borough ZIP Code areas were divided into those below and above an age range benchmark for weathering. The COVID markers, other health markers, and SE factors of the below benchmark ZIP Code areas were compared with those of the above benchmark areas. We also regressed premature mortality rates of the areas within each borough against the COVID markers to reveal any associations.

We have used only simple, basic statistical analytical techniques for pattern revelation: bivariate and multivariate regression, t-tests, and Mann–Whitney tests. See this chapter's Introduction for explanation of comparing Brooklyn with Queens and Manhattan with the Bronx.

2.3 Results

2.3.1 Brooklyn and Queens

2.3.1.1 Percent Positive Swab Tests

The ZIP Code areas of Brooklyn and Queens show no significant difference in their percent positive swab tests (Table 2.1).

Percent of the population that is foreign-born associates significantly with percent positive swab tests for Brooklyn but not for Queens; percent housing

Table 2.1 Statistical
descriptions of CoViD
markers: Brooklyn and
Queens

	Brooklyn	Queens	P
Percent positive			
Mean	29.41	30.48	NS
Median	30	32	NS
SD	4.85	4.74	
Minimum	18	19	
Maximum	37	39	
Cases per 100,000			
Mean	2013.46	2575.4	0.0003
Median	2056	2608	0.0002
SD	645.94	736.06	
Minimum	877	1328	
Maximum	4068	4306	
Deaths per 100,000			
Mean	201.54	206.7	NS
Median	174	189	NS
SD	102	98	
Minimum	60	42	
Maximum	620	475	

units with extreme overcrowding, percent Latinx, and percent over 65 weakly
but significantly associated with percent positive swab tests in Queens but not in
Brooklyn (Table 2.2).

The multivariate regressions with SE factors as independent variables and percent
positive as dependent variable also produced different results. In Brooklyn, only
race (Black %), ethnicity (Latinx %), and origins (foreign-born %) survived the
regression (R-square = 0.72). In Queens, many aspects of the SE system ended up in
the resulting equation: race (Black %), ethnicity (Latinx %), educational attainment
(college %), age (% over 65), and economic (rent stress) and produced an R-square
of 0.69. Brooklyn and Queens showed 12–15 times as many positive swab tests as
confirmed cases (Table 2.1).

2.3.1.2 Cases/100,000

However, cases per unit population in Queens outnumber those for Brooklyn
significantly on average and on median by 27–28%. Associations between case
rates and SE factors differ greatly from those between percent positive swabs and
SE factors (Table 2.3). In Brooklyn, such non-race/ethnicity factors as percent
college, median income, and rent stress show much higher R-squares than in
Queens, a reversal of that picture for percent positive. The equations produced
by the multivariate regressions also show a reversal: in Brooklyn, an array of SE
factors (race, age, and household economics) are the independent variables whereas
for Queens, only race/ethnicity factors are. The bottom half of Table 2.3 shows
that the two boroughs are two separate systems. The multivariate regression for

Table 2.2 Percent positive and associations with SE factors

SE factor	Brooklyn			Queens		
	R-sq	P	pos/neg	R-sq	P	pos/neg
Black %	0.2696	0.0006	pos	0.1314	0.0035	pos
College %	0.4885	<0.0001	neg	0.5842	<0.0001	neg
eoc %	NA			0.066	0.0314	pos
Foreign %	0.2494	0.001	pos			
Med. Inc.	0.384	<0.0001	neg	0.4663	<0.0001	neg
Latinx %	NA			0.0796	0.0199	pos
Over 65 %	NA			0.0536	0.0475	pos
Rent stress	0.1419	0.0124	pos	0.3049	<0.0001	pos
White %	0.4646	<0.0001	neg	0.2941	<0.0001	neg

SE	Brooklyn and Queens		
	R-sq	P	pos/neg
Black %	0.162	<0.0001	pos
College %	0.5407	<0.0001	neg
eoc %	0.0329	0.0451	pos
Foreign %	0.1238	0.0003	pos
Med. Inc.	0.3733	<0.0001	neg
Latinx %	0.075	0.0046	pos
Rent stress	0.2425	<0.0001	pos
Unemploy	0.0396	0.0311	pos
White %	0.3683	<0.0001	neg

Brooklyn positive % = 14.89+0.1 (black%) + 24.135 (foreign%) − 0.1365 (Latinx%)
R-sq = 0.72
Queens positive % = 20.019 + 0.062 (black %) − 0.19 (college %) + 0.09 (Latinx %) + 0.306 (over65 %) + 0.198 (rent stress)
R-sq = 0.69
BklnQns positive % = 42.16 − 0.002 (med.inc.) + 0.08 (Latinx %) − 0.166 (poverty) − 0.09 (white %)
R-sq = 0.60

Brooklyn plus Queens produces an R-square less than those of the two separate boroughs. None of the SE factors associated with case rate achieved R-square above 0.4 (Table 2.3).

This holds true for Brooklyn, Queens, and the combined Brooklyn/Queens Zip Code areas. Associations in Brooklyn are stronger than those in Queens, and the multivariate regressions also produced a much higher R-square for Brooklyn (0.64) than for Queens (0.40). The multivariate regression for the combined Brooklyn/Queens Zip Code areas also produced an R-square of only 0.40 and showed that the two boroughs do not form a single system for case rate.

The Brooklyn equation from the multivariate regression includes two measures of household economic distress, poverty rate and rent stress, as well as race and age factors. The Queens equation includes only race and ethnicity factors.

Table 2.3 Case rates and associations

SE factor	Brooklyn			Queens		
	Cases/10E5 associated with SE factors					
	R-sq	P	pos/neg	R-sq	P	pos/neg
Asian %	NA			0.0897	0.0142	neg
Black %	0.0705	0.0616	pos	0.1076	0.0078	pos
College %	0.3704	<0.0001	neg	0.2469	0.0001	neg
% EOC	NA			0.0713	0.0262	pos
Foreign %	0.1475	0.0109	pos		NA	
Latin %	NA			0.1247	0.0044	pos
Med. Inc.	0.3669	<0.0001	neg	0.1154	0.006	neg
Over 65 %	0.2898	0.0003	pos	NA		
Poverty	0.0938	0.0365	pos		NA	
Rent stress %	0.2068	0.0027	pos	0.0476	0.0581	pos
White %	NA			0.0938	0.0124	neg

Brooklyn and Queens together

SE factor	R-sq	P	pos/neg
Black %	0.423	0.0269	pos
College %	0.3025	<0.0001	neg
Med.inc.	0.1476	0.0001	neg
Latinx %	0.045	0.0232	pos
Over65%	0.0283	0.0582	pos
Rent stress	0.102	0.001	pos
White %	0.0953	0.0015	neg

Bkln cases/10E5 $= -918 + 5.986$ (black %) $+ 82.35$ (over 65 %) $+ 23.39$ (poverty) $+ 27.5$ (rent stress)

R-sq $= 0.64$

Qns cases/10E5 $= 2925.59 - 14.14$ (Asian%) $+ 22.36$ (Latinx %) $- 13.84$ (white %)

R-sq $= 0.40$

Brooklyn and Queens together

Cases/10E5 $= 2163.01 - 8.87$ (Asian %) $- 29.28$ (college %) $+ 14.18$ (Latinx %) $+ 72.99$ (over 65 %)

R-sq $= 0.43$

2.3.1.3 COVID Mortality Rates Per 100,000

No significant difference was found between the COVID mortality rate means and medians of Brooklyn and Queens (Table 2.1).

Although most SE factors show weak to moderate associations with COVID death rates in Brooklyn and Queens (Table 2.4), the multivariate regression yielded an equation that 'explains' 81% of the pattern of death rates over the ZIP code areas of Brooklyn. This equation includes poverty rate, percent white, and percent over age 65. Poverty rate, the highest ranked SE factor for R-square in Queens, swamped all other SE factors in the multivariate regression. Thus only 20% of the pattern of death rates over the ZIP code areas of Queens is 'explained'.

Table 2.4 COVID death rates and associations with SE factors

SE factor	Brooklyn			Queens		
	R-sq	P	pos/neg	R-sq	P	pos/neg
Black %	0.1115	0.0246	pos	NA		
College %	0.1827	0.0048	neg	0.0939	0.0124	neg
eoc %	NA			0.1819	0.0006	pos
Foreign %	0.12	0.0203	pos	NA		
Latinx %	NA			0.0936	0.0125	pos
Med. Inc.	0.2294	0.0016	neg	0.1225	0.0047	neg
Over 65 %	0.5673	<0.0001	pos	NA		
Poverty rate	NA			0.2022	0.0003	pos
White %	0.0714	0.0603	neg	NA		

Brooklyn and Queens together

SE factor	R-sq	P	pos/neg
Black %	0.0289	0.0564	pos
College %	0.1399	0.0001	neg
Foreign%	0.0421	0.0272	pos
Med. Inc.	0.1696	<0.0001	neg
Over 65%	0.0789	0.0037	pos
Poverty	0.0633	0.0086	pos
White %	0.0313	0.0493	neg

Brooklyn deaths/10E5 $= -69.79 + 18.89$ (%over 65) $+ 4.063$ (poverty) $- 1.44$ (white %)
R-sq $= 0.81$
Queens deaths/10E5: poverty rate swamps all others in multivariate regression
R-sq $= 0.20$
Deaths/10E5 $= 148.54 - 1.544$ (Asian %) $- 0.0041$ (med.inc.) $+ 1.977$ (Latinx %) $+ 12.167$ (over65 %)
R-sq $= 0.38$

The combination of Brooklyn and Queens ZIP Codes areas produced no associations between SE factors and death rate with R-Square above 0.2. The multivariate regression yielded an R-square of 0.38, much below that for Brooklyn alone. Furthermore, the resulting equation did not include poverty rate, a factor in the Brooklyn equation and the sole survivor of the regression process for Queens. As with percent positive and case rate, the two boroughs operate as separate systems for COVID mortality rate.

2.3.1.4   COVID Markers and Other Public Health Indicators

The previous chapter analyzed patterns of premature mortality rate at the community district level. In regressing premature ZIP Code area mortality rate with COVID markers (percent positive, case rate, and death rate, we again see that Brooklyn and Queens function differently (Table 2.5).

Table 2.5 Premature Mortality and COVID Markers: Brooklyn and Queens

Associations between premature mortality and COVID markers		
	Brooklyn	Queens
CoVid Marker	R-sq	R-sq
Percent positive	0.1962	NA
Case rate	0.1209	NA
Death rate	0.4181	0.2069
Brooklyn		
Significant difference in percent positive mean: $P = 0.0350$		
Significant difference in percent positive median: $P = 0.0393$		
No significant difference in standard deviation		
No significant difference for case rate mean or median		
Significant difference for case rate standard deviation: $P = 0.0366$		
No significant difference in death rate mean		
Significant difference in death rate median: $P = 0.0273$		
Trend to difference in death rate standard deviation: $P = 0.0962$		
Queens		
Significant difference in percent positive mean: $P = 0.0183$		
Trend to difference in percent positive median: $P = 0.0626$		
Significant difference in percent positive standard deviation: $P = 0.01$		
Significant difference in case rate mean: $P = 0.0463$		
Significant difference in case rate median: $P = 0.0468$		
No significant difference in case rate standard deviation		
Significant difference in death rate mean: $P = 0.0032$		
Significant difference in death rate median: $P = 0.0027$		
No significant difference in death rate standard deviation		

Contrast in COVID Markers above and below 45–49 age beginning of weathering

All three markers show significant association with premature mortality rate over the Brooklyn ZIP Code areas but only COVID death rate does in Queens. On the other hand, when we divide the ZIP Code areas into those with weathering that begins below the 45–49 age range and those that begin at or above (Table 2.5), we see significant differences and one trend to difference for the means and medians of COVID markers in Queens. Brooklyn also shows several significant differences.

The public health markers (heart, cancer, flu/pneumonia, and diabetes mortality rates) show no difference between the ZIP Code areas in either Brooklyn or Queens below and above the weathering benchmark of 45–49 years of age with the exception of diabetes mortality rate. Brooklyn, in particular, shows a large difference in mean (28.95 vs. 19.61) and median (28.9 vs. 16.4) diabetes mortality rates for areas with below and above the benchmark. When premature mortality rate is regressed against the public health markers, all regressions resulted in significant associations in both boroughs but diabetes mortality rate ranked highest by far for R-square (0.66 in Brooklyn and 0.28 in Queens). Furthermore, in the multivariate regression, diabetes mortality and heart disease mortality together associate with 77% of the pattern of COVID mortality rates over the ZIP Codes of Brooklyn; diabetes mortality and flu/pneumonia mortality together associate with 28% of the COVID mortality rate pattern over the ZIP Codes of Queens.

Diabetes as a factor in premature mortality, 'weathering', and COVID mortality patterns makes sense. Some ZIP Codes begin 'weathering' as young as the 25–30 age range. Type 2 diabetes cases began showing up in large numbers in children as young as eleven toward the beginning of the obesity epidemic (Kao and Sabin 2016). Type 1 diabetes ('childhood' diabetes) has also increased in incidence with the obesity epidemic (Mitchell 2017).

The maps of Brooklyn COVID death rates and premature mortality rates (not shown) reveal a swath of elevated premature and COVID mortality rates along the poverty belt that runs from Bedford-Stuyvesant southeast into Brownsville, East New York, and Starrett City. Coney Island also had very high COVID and premature mortality rates. There were some anomalies: Borough Park and Bensonhurst had low premature mortality rates but moderately high (200–250/100,000) COVID mortality rates.

With the exception of the Rockaway Peninsula, Queens ZIP Code areas enjoyed low premature mortality rates, mostly 100–150/100,000 (map not shown). None of the areas, including the Rockaway Peninsula suffered from rates of over 300/100,000 premature deaths. Only eight out of the 56 areas had rates above 150/100,000. The geography of COVID death rates contrasts strongly with that of premature death rates. Only 16 areas had COVID mortality rates below 150; 21 had mortality rates over 200, including ten with rates over 250. Besides the Rockaway Peninsula, the five areas immediately south and southeast of LaGuardia Airport suffered COVID mortality rates above 300/100,000. The map of COVID mortality rates for Queens showed no resemblance to that of premature mortality with the exception of the Rockaway Peninsula.

Of the ten Queens areas with COVID mortality above 300/100,000, six have over 50% of their population foreign born. A seventh has a very high proportion of elderly (23% over 65). The three others are on the Rockaway Peninsula. Of the five areas with COVID mortality below 100, none have over 50% of their populations foreign-born.

2.3.2   Manhattan and the Bronx

2.3.2.1   Overview

Although the populations of Manhattan and the Bronx don't rival those of Brooklyn and Queens, their areas are very small and produce much higher population densities than reside in Brooklyn and Queens. New York City's major business districts, cultural institutions, and governmental offices lie in Manhattan. Yet, Manhattan's COVID burden is lower than that of the Bronx (Table 2.6), Brooklyn or Queens, by

Table 2.6 Comparison of Manhattan and the Bronx for COVID-19 Burden

	Manhattan	Bronx	P	t
Percent positive swab samples				
Average	19.15	30.58	<E-11	9.64
Median	17	31	9.79E-10	
SD	5.32	2.8	0.002	
Minimum	11	23		
Maximum	28	35		
Confirmed cases per 100,000				
Average	1465	3126.58	<E-11	11.79
Median	1356	3040.5	9.27E-05	
SD	588.42	434.25		
Minimum	520	2402		
Maximum	2719	4137		
Confirmed deaths per 100,000				
Average	137.49	236.54	2.39E-05	4.6
Median	117	222	5.27E-05	
SD	82.6	79.07		
Minimum	0[a]	99		
Maximum	325	449		

[a]NYC Dept. of Health reported no deaths in Battery Park City
This report does not appear credible

any of the three markers (percent positive, confirmed cases per 100,000 or confirmed deaths per 100,000). Although the percent positive swab tests for the Bronx show similar means and medians to those of Brooklyn and Queens, the confirmed case rate and confirmed death rate means and medians exceed those of Brooklyn and Queens. Indeed, the mean and median case rate toted up above 3000 per 100,000 population on May 31, 2020, the date of our cumulative COVID markers. The death rate of the Bronx is about 10–15% higher on average and median that those of Brooklyn and Queens, shown on Table 2.1. This section of Chap. 2, thus, examines the highest and lowest burdened boroughs with respect to COVID. Chapter 3 will show that this section examines the highest and lowest counties of the 24-county metropolitan region with respect to COVID mortality rate.

2.3.2.2 Percent Positive

Most SE factors that associate with Manhattan's percent positive swab tests do so with R-squares above 0.5 (Table 2.7).

These factors include race/ethnicity, educational attainment, household economics, and unemployment rate. In the Bronx, only percent white and percent college or higher degree achieve an R-square over 0.5. The SE factors that associate with percent positive in both boroughs do so in the same direction. For example:

Table 2.7 SE associations with percent positive swabs

SE Factor	Manhattan			Bronx		
	R-sq	*P*	pos/neg	R-sq	*P*	pos/neg
Black %	0.3586	0.0001	pos	0.2095	0.0142	pos
Latinx %	0.629	<0.0001	pos	0.1594	0.0303	pos
White %	0.7337	<0.0001	neg	0.5919	<0.0001	neg
College %	0.7632	<0.0001	neg	0.5333	<0.0001	neg
Extreme overcrowding	0.0807	0.0543	pos	NA		
Individual median income	0.8333	<0.0001	neg	0.463	0.0002	neg
Over 65 yrs %	0.1024	0.0344	neg	0.2933	0.0037	neg
Poverty rate	0.7056	<0.0001	pos	0.2143	0.0132	pos
Rent stress	0.634	<0.0001	pos	0.4098	0.0005	pos
Unemployment	9.5499	<0.0001	pos	NA		

Manhattan % positives $= 25.89 - 0.00015$ (median income) $+ 0.059$ (Latinx %)
R-sq $= 0.85$
Bronx % positives $= 13.14 - 0.3$ (college %) $+ 0.34$ (rent stres) $+ 0.61$ (over 65 yrs %)
R-sq $= 0.74$
Bx % positives $= 23.745 + 0.063$ (black %) $- 0.115$ (college %) $+ 0.15$ (rent stress)
R-sq $= 0.68$

percent Black positively associates with percent positive in both boroughs and percent college negatively. The multivariate regression of SE factors with percent positive shows that individual median income and percent Latinx vary with percent positive over the ZIP code areas of Manhattan 85%. Two possible results of multivariate regression arise for the Bronx: percent college, rent stress, and % over 65 yield an R-square of 0.74 whereas percent black, percent college, and rent stress produce an R-square of 0.68.

2.3.2.3 Case Rates Per 100,000

The bivariate regression results for cases/100,000 are wild (Table 2.8).

For many SE factors, the results in Manhattan and the Bronx differ in direction. Percent Latinx associates positively in Manhattan and negatively in the Bronx; percent college, negatively in Manhattan and positively in the Bronx. Of the nine Manhattan associations, six have R-squares over 0.5. Poverty rate is the only SE factor in the Bronx with an R-square above 0.4. Indeed, poverty rate swamps all other SE factors in the multivariate regression for the Bronx, whereas Black percent and Latinx percent account for 73% of the case rate variability over the ZIP Code areas of Manhattan.

Table 2.8 SE associations with COVID cases per 100,000

SE factor	Manhattan			Bronx		
	R-sq	*P*	pos/neg	R-sq	*P*	pos/neg
Asian %	0.1495	0.0125	neg	NA		
Black %	0.3649	0.0001	pos	NA		
Latinx %	0.5568	<0.0001	pos	0.366	0.001	neg
White %	0.5886	<0.0001	neg	NA		
College %	0.5955	<0.0001	neg	0.2029	0.0157	pos
Extreme overcrowding	NA			0.2927	0.0037	neg
Median income	0.6359	<0.0001	neg	0.3186	0.0024	pos
Over 65 years %	NA			0.3806	0.0008	pos
Poverty rate	0.5852	<0.0001	pos	0.4108	0.0004	neg
Rent stress	0.5659	<0.0001	pos	0.1439	0.0381	neg
Unemployment	0.4841	<0.0001	pos	NA		

Manhattan case rate = 824.159 + 12.74 (black %) + 18.21 (Latinx %)
R-sq = 0.73
Bronx case rate: poverty rate swamps all others in regression
R-sq = 0.41

Table 2.9 SE associations with COVID deaths per 100,000

SE factor	Manhattan			Bronx		
	R-sq	*P*	pos/neg	R-sq	*P*	pos/neg
Black %	0.3342	0.0002	pos	NA		
Latinx %	0.3897	<0.0001	pos	NA		
White %	0.5502	<0.0001	neg	NA		
College %	0.5365	<0.0001	neg	NA		
Median income	0.6034	<0.0001	neg	NA		
Poverty rate	0.5436	<0.0001	pos	NA		
Rent stress	0.4739	<0.0001	pos	NA		
Unemployment	0.4307	<0.0001	pos	NA		

For Manhattan, median income swamped all other factors in multivariate regression
No regression was possible for the Bronx

2.3.2.4 COVID Deaths Per 100,000

Bivariate regression yields unexpected results for COVID deaths/100,000 (Table 2.9).

For Manhattan, eight SE factors significantly associate with death rate, four with R-squares over 0.5 and all with R-squares over 0.3. None of the SE factors in the dataset associate with death rate in the Bronx although percent housing units with extreme overcrowding has a weak trend with R-square 0.0798 ($p = 0.0976$).

2.3.2.5 Differing Public Health Cohesion in Manhattan and the Bronx

For the Bronx, no association exists between percent positive swabs and case rate or percent positive swabs and death rate. Case rate and death rate associate with an R-square of 0.43. For Manhattan, percent positive associates with case rate with an R-square of 0.80 and with death rate with an R-square of 0.74. Case rate associates with death rate in Manhattan with an R-square of 0.86. Thus, in Manhattan, the markers of COVID present a cohesive picture but not in the Bronx.

Premature death rates in the Bronx significantly exceed those in Manhattan on average (Bronx = 224, Manhattan = 161) and on median (Bronx = 222, Manhattan = 129). Although the Bronx also has significantly higher COVID markers than Manhattan, its COVID markers show no association with its premature death rates over its ZIP Code areas; there is no relationship, not even a weak trend to association. On the other hand, the Manhattan COVID markers each associate significantly with premature death rates of the ZIP Code areas: percent positive R-sq = 0.35, case rate R-sq = 0.32, death rate R-sq = 0.50 (Figs. 2.1 and 2.2).

Large swaths of Manhattan had low rates of both premature mortality and of COVID mortality. East Harlem presented elevated rates of premature mortality

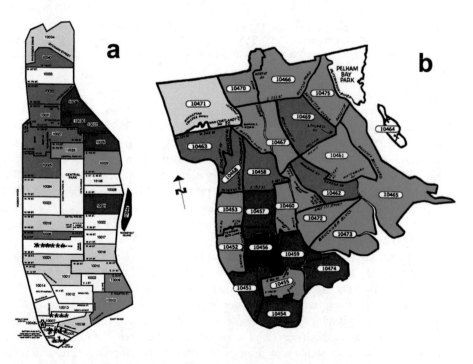

Fig. 2.1 Premature Mortality Rate by ZIP Code Area of Manhattan (**a**) and the Bronx (**b**). Black = above 300, red = 250–300, green = 200–250, orange = 150–200, yellow = 100–150, white = below 100. The white areas with an asterisk have fewer than 10,000 residents and are not part of the analysis

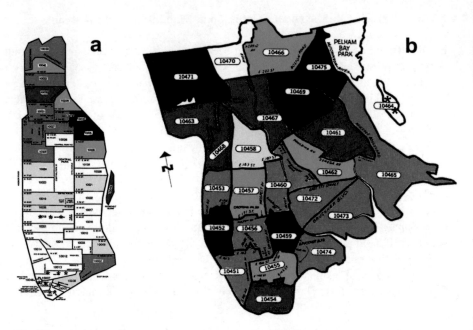

Fig. 2.2 COVID Mortality Rate by ZIP Code Areas of Manhattan (**a**) and the Bronx (**b**). Black = above 300, red = 250–300, green = 200–250, orange = 150–200, yellow = 100–150, and white = below 100. The white areas with a cross mark have fewer than 10,000 residents and are not part of the analysis

in four areas and of COVID mortality in two. Roosevelt Island's rates were also elevated. Eighteen of the 35 ZIP Code areas of Manhattan had premature mortality rates below 150; twenty had COVID mortality rates below 150. Twelve Bronx areas had premature mortality rates above 250 and only one below 150. Ten Bronx areas (out of 24) had COVID mortality rates above 250 and only two below 150. The southwest Bronx had a 6-area swath of high premature mortality. Although COVID mortality rates were high in the southwest Bronx, the northern Bronx presented a large swath of elevated COVID mortality rates: seven of the ten areas with rates above 250 contributed to the Northern swath.

When the ZIP Codes are divided into those below and those above the weathering benchmark (45–49 for Manhattan and 40–44 for the Bronx), Manhattan shows significant differences for all three COVID markers (Table 2.10).

The Bronx shows significant difference in percent positive swab tests but not for either case rate or death rate. When these ZIP Code areas are compared for the SE factors in our dataset, both boroughs show significant differences for many factors (Table 2.11), in the expected direction. Poverty rate in both boroughs, for example, is higher on average and median in the ZIP Code areas with weathering beginning below the benchmark than in those beginning above it.

It is obvious, however, that the Bronx presents a much worse SE profile than Manhattan, whether one looks at poverty rate, rent stress, percent college, or median

Table 2.10 COVID Markers of areas below and above weathering benchmark. Manhattan benchmark: 45–49 years of age. Bronx benchmark: 40–44 years of age

	Manhattan		Bronx	
	Mean	Median	Mean	Median
% positive				
Below	22	23.5	31.9	32
Above	16.4	15	28.4	30
P	0.0009	0.0027	0.0017	0.0073
Case rate				
Below	1732.4	1633.5	3049	2983
Above	1223.4	1076	3256	3155
P	0.0076	0.0061	NS	NS
Death rate				
Below	175.6	168.5	240	228
Above	103.2	81.5	231	211
P	0.0066	0.0033	NS	NS

Number of Mn areas with weathering at 25–29: 2
Number of Mn areas with weathering over 50: 4
Total number Mn areas: 35
Number of Bx areas with weathering at 25–29: 4
Number of Bx areas with weathering over 50: 0
2 Bx areas with weathering at 45–49
Total number Bx areas: 24

income. When diabetes mortality rate is compared between the below and above benchmark areas, it is significantly different in Manhattan but not in the Bronx. In fact, it appears nearly identical on average and on median for the below and above benchmark ZIP Code areas of the Bronx.

2.3.3 Public Health of the Four Boroughs

The relationships between diabetes mortality rate and the three COVID markers differ among the boroughs (Table 2.12).

In Manhattan, even percent positive shows fairly strong association with diabetes mortality, and that of COVID mortality rate has an R-sq above 0.5. In Brooklyn, the relationship tightens from percent positive to case rates and then to death rates, with the last having an R-sq over 0.6. In Queens, the three associations are rather weak. In the Bronx, percent positive has no association with diabetes mortality rate and case rate and death rate have weak ones. When we regressed premature mortality rate against the cause-specific mortality rates in our dataset (Table 2.12), differences between the boroughs also appeared. Both Brooklyn and Manhattan had associations between premature mortality rate and diabetes mortality rate with R-squares above 0.6, but cancer mortality had no association with premature mortality in Manhattan or the Bronx. It did have moderately strong associations in Brooklyn

Table 2.11 Comparison of SE factors: below and above weathering benchmark

SE factor	Manhattan					
	Below		Above			
	Mean	Median	Mean	Median	Mean p	Median p
Asian %	9.07	6.20	16.77	16.10	0.0210	0.0061
Black %	26.93	16.44	5.38	4.80	0.0010	0.0011
College %	47.62	37.68	72.69	77.98	0.0002	0.0009
Med. Inc.	38,764	29,173	70,118	74,580	0.00006	0.0004
Latinx %	33.36	25.46	13.99	10.95	0.0040	0.0004
Poverty	22.73	25.43	19.68	10.46	4.50E-05	0.0004
Rent stress	36.94	37.36	33.7	32.08	0.0362	0.0136
Unemploy	6.47	5.76	3.90	3.90	0.0025	0.0040
White %	42.53	32.36	69.72	74.20	0.0004	0.0013
Bronx						
College %	14.7	13.41	28.78	26.6	0.0005	0.0016
EOC %	5.89	6.36	3.17	3.48	0.0014	0.0035
Foreign %	38	40	28	28	0.0002	0.0016
Med. inc.	18,628	17,148	30,159	30,422	0.0008	0.0029
Latinx %	60.6	68.15	43.29	45.7	0.0165	0.0235
Poverty	34.48	36.11	17.51	14.66	0.0001	0.0019
Rent stress	50.58	51.38	40.11	40.24	9.00E-06	0.0003
White %	13.24	11.84	40.03	43.24	0.00003	0.0004
Diabetes mortality						
MN	22.94	22.94	12.85	9.44	0.0334	0.0068
BX	23.02	22.82	22.82	23.34	NS	NS

NS not significant, *EOC* extremely overcrowded housing, *foreign* foreign-born, *Med. Inc.*, individual median income

Table 2.12 Associations of diabetes mortality rates with markers of COVID: four boroughs

Borough	R-sq for % positive	R-sq for case rate	R-sq for death rate
Brooklyn	0.1718	0.2408	0.6114
Queens	0.1466	0.0717	0.1551
Manhattan	0.4173	0.3470	0.5257
Bronx	NA	0.1516	0.1252

Associations of premature mortality rate with specific rates

Borough	R-sq for diabetes	R-sq for cancer	R-sq for flu/pneumonia	R-sq for heart
Brooklyn	0.6628	0.3935	0.1096	0.1075
Queens	0.2791	0.2035	0.1696	0.1613
Manhattan	0.6349	NA	0.3505	0.2902
Bronx	NA	NA	NA	0.3960 (neg)

NA no association

All associations with premature mortality are positive but Bronx heart and Queens. Indeed, the Bronx had only one association: heart disease mortality negatively associated with premature mortality.

From all these regressions and comparisons, we can say that Brooklyn and Manhattan show strong systems with many close relationships and deep differences between ZIP Code areas with early and late beginning of weathering. Queens shows a rather weak system with weak to moderate relationships and fewer significant differences in SE factors between areas below and areas above the weathering benchmark in comparison to Brooklyn. Although the Bronx showed many significant differences between SE factors of the areas below and above the weathering benchmark (Table 2.11), most means and medians revealed a poor SE profile. For example, the above weathering Bronx areas had an average median income of $30,159 whereas the analogous Manhattan areas had one of $70,189. Rent stress in the below weathering Bronx areas averaged 50.58% but in the analogous Manhattan areas, 36.94%.

The health outcomes in the Bronx showed few connections to each other. Premature mortality had no association with diabetes, cancer, or flu/pneumonia mortality rates and a negative one with heart disease mortality rate. Premature mortality rate had no association in the Bronx with any of the three COVID markers, although premature mortality rate and all three COVID markers on average and median exceeded those of the other three boroughs. Health in the Bronx was so poor that the very benchmark for weathering had to be notched down from the 45–49 age range of the other three boroughs to 40–44.

The Bronx showed a few weak associations between health markers, in contrast to the other boroughs. For example: Diabetes, flu/pneumonia, and cancer mortality rates had no association with premature mortality rate. For Manhattan and Brooklyn, diabetes and premature mortality rates associated with R-squares above 0.6, and for Queens, about 0.3. When the SE factors are regressed against each other, a different picture emerges. They do form a system in the Bronx. Table 2.13 uses percent college or higher degree and rent stress as two examples.

Most SE factors in the dataset significantly associate with percent college or higher degree. Five associations had R-squares above 0.5 with median income ranking highest at almost 0.9. Similarly, most SE factors associated significantly with rent stress. Six had R-squares over 0.4 with percent foreign born ranking highest at 0.68. The Bronx clearly has a coherent socioeconomic geography but not a public health one.

2.4   Discussion

Allostatic load, an index system created by Bruce McEwen, measures the health consequences of stress, especially chronic stress. It is a multifactorial index with factors ranging in number between six and 24, depending on the research team and its objectives (McEwen 2015). Allostasis is the attempt by the body to restore

Table 2.13 Two examples of Bronx SE factor relationships: % college and rent stress

SE Factor	R-sq	P	pos/neg
College or higher degree percent			
EOC %	0.2258	0.0110	neg
Foreign %	0.1775	0.0231	neg
Asian %	0.1752	0.0239	pos
Med. Inc.	0.8870	<0.0001	pos
Latinx %	0.5634	<0.0001	neg
% over 65	0.7222	<0.0001	pos
Poverty	0.6832	<0.0001	neg
Rent stress	0.3721	0.0009	neg
Unemploy	0.1401	0.0403	neg
White %	0.6386	<0.0001	pos
Rent stress			
EOC %	0.4783	0.0001	pos
Foreign %	0.6753	<0.0001	pos
Med inc.	0.5348	<0.0001	neg
Latinx %	0.3535	0.0007	pos
Over 65 %	0.6362	<0.0001	neg
Poverty	0.4695	0.0001	pos
Unemploy	0.1830	0.0210	pos
White %	0.4805	0.0001	neg

No multivariate regression because of high med. inc. R-sq

Rent stress $= 39.546 + 46.19$ (foreign %) $- 0.674$ (% over 65)

R-sq $= 0.84$

homeostasis during and after a stressful event that required coping mechanisms beyond the normal ranges of physiological processes (Romeo and McEwen 2006). Repetitive or chronic stress that elicits stress hormones on a long-term basis results in allostatic overload. Allostatic overload is measured on individuals by neurophysiologic, metabolic, circulatory, psychological, and other markers.

Various researchers group individuals according to socioeconomic measures such as household income, educational attainment, and occupational grade and analyze the allostatic loads of the groups to reveal the effects of these individual socioeconomic factors on health (Kim et al. 2018). Other researchers measure allostatic loads of residents of neighborhoods with differing environmental and SE conditions to reveal the effects of community conditions and structure on health (Tan et al. 2017). Both these research models depend on individual health outcomes.

The analyses in this chapter show the existence of population allostatic overload, above and beyond individual and small group dynamics. The boroughs exhibit a spectrum of public health response to SE factors from very connected (Manhattan) to fairly connected (Brooklyn) to weakly connected (Queens) to unconnected (Bronx). They also show differences in the coherence of health outcomes with

Manhattan having strong associations between the health outcomes in our dataset and the Bronx having a very few weak associations. Brooklyn shows slightly weaker associations than Manhattan and Queens, weaker than Brooklyn.

The disconnection between SE factors and health outcomes and between health outcomes in the Bronx indicates a breaching of a threshold for population allostatic load. If the Bronx had low or even moderate health indicators, this disconnect would hint at extreme resilience; but that is not the case. The Bronx has the highest COVID case rates and death rates besides very high premature mortality rates over a high percent of its ZIP code areas. The disconnect in this context presents the possibility that the Bronx suffered a regime change toward extreme vulnerability and chronically high rates of public health problems that should be considered borough-wide population allostatic overload. Although the SE factors in the Bronx associated with each other and form a system, the lack of connection with the health markers may mean that the SE system is too weak and resource-drained to budge the health markers. The SE system may have declined below the threshold of health effectiveness.

The geography of COVID mortality in the Bronx also hints that the healthcare facilities (both nursing homes and hospitals) seeded COVID across the North Bronx. ZIP code area 10471 had a low premature mortality rate but a very high COVID mortality rate. It contains a large number of large nursing homes. Montefiore Medical Center, Albert Einstein College of Medicine, and several smaller hospitals are strung across the North Bronx. The Manhattan COVID mortality geography also hint that hospitals seeded COVID in their ZIP Code areas: 10032 hosts New York-Presbyterian Medical Center (formerly Columbia Presbyterian) and 10021 is often referred to as BedPan Alley because of the numerous healthcare facilities there such as Rockefeller University Hospital, Memorial Sloan Kettering Cancer Center, and Hospital for Special Surgery. These two ZIP Code areas show higher COVID mortality rates than the surrounding areas and higher than expected for their SE and premature mortality contexts.

Besides the Bronx, Queens showed high levels of COVID markers with higher case rates and death rates than Brooklyn, although Brooklyn has slightly more residents and higher population densities. Brooklyn also suffered higher premature mortality rates than Queens. Premature mortality rates and COVID mortality rates in Queens diverged massively in geography, with the exception of the poverty belt of the Rockaway peninsula. The huge majority of ZIP Code areas of Queens enjoy premature mortality rates below 150/100,000. In contrast, eleven areas had COVID mortality rates above 250, ten of them above 300. Three of the ten were on the Rockaway Peninsula. Of the others, five clustered about LaGuardia Airport and had over 50% of their populations foreign-born.

Many immigrant populations mistrust the authorities, for good reason in this era of Trump's anti-immigrant policies. Furthermore, language barriers prevent rapid, efficient information flow, especially if the language isn't one of the major ones spoken in New York such as Spanish, Chinese, or Russian. If members of the immigrant community who also speak English and can serve as conduits of information trust the information suppliers as having the community's wellbeing as a priority, the language problem diminishes. If the health authorities have not established a long-term relationship with the immigrant communities, attempting to convey vital information to them in an emergency will fail. The contrasting geography of premature mortality and COVID mortality rates document that failure.

In a discussion on the American Public Health Association's 'Spirit of 1848' caucus listserv, the possibility that Queens immigrant communities were not receiving prevention information came up. NYC Health Department staff and their friends insisted that the department was energetically pursuing outreach to these communities and getting information to the houses of worship, social clubs, and local clinics. This discussion occurred in late March as the case load was building to its crest in the city and Queens stood out from Brooklyn in case rates.

Apparently, disaster preparedness did not include long-term communication between the myriad diverse communities of the city and the health department. Queens happens to be the most diverse borough with more languages spoken in such neighborhoods as Astoria, Flushing, and East Elmhurst than in any other borough's neighborhoods. If you want to hear Judeo-Tajik, just go to the King David Restaurant (after the pandemic) in Forest Hills. They sometimes have live Bukharan music as well. This level of diversity requires a nearly block-by-block continuous communication between civic entities and the residents so that rapid adaptations can occur in emergencies. Not putting the effort into such communication is sheer negligence and deficient disaster-readiness.

Even some long-established ethnic enclaves need this continuous two-way communication. ZIP Codes areas 11219 and 11230 in Brooklyn have large Chasidic and Italian populations, respectively. These cultures are extremely social. Masks, social distancing, home lockdown, and other barriers to close contact in large groups are alien and unwelcome requirements. For these populations, the communication has been one-way, namely the community representatives tell aspiring office-seekers what they want in return for political endorsements.

The DoH has had serious problems with risk behaviors in these communities but has never entered long-term continuous relationships with them. In the COVID emergency, this lack has shown up in the higher-than-expected COVID mortality rates in Borough Park and Bensonhurst (11219 and 11230 respectively). Mayor de Blasio personally broke up a large Chasidic funeral with a huge show of anger at the flouting of the rule against mass events (Hasidic Funeral 2020). Yet all along, he had allowed much other substandard behavior in that community such as failure of the schools to meet educational standards (Yeshiva Investigation 2019) and several public health-damaging practices (Friedman 2019). Suddenly, he demanded obedience to civil society's requirements.

Poor New York City neighborhoods of color incurred damaging public policies for many decades from redlining, urban renewal, and planned shrinkage to present day encouragement of gentrification. In most instances, the city agencies attempted to target the damage to the poor neighborhoods. Thus, particular neighborhoods in Manhattan such as Harlem and Chinatown, in Brooklyn such as Bedford-Stuyvesant and Brownsville, and in Queens such as the Rockaway Peninsula and Jamaica lost housing, essential services, and proper supervision of police. We have described how housing was destroyed because of closing of fire companies as part of planned shrinkage and how the populations of these neighborhoods were greatly reduced in the late 1970s. The destabilizing of communities cascaded into a public health and safety disaster that included a TB epidemic, acceleration of the AIDS epidemic, upsurges in drug deaths, violent crime, and low-weight births (Wallace and Wallace 1998). The maps of premature mortality reflect continued health consequences and vulnerability to further impacts such as the 2007–2008 financial/housing crisis (The Great Recession). With the exception of the high COVID mortality ring around LaGuardia Airport, the COVID mortality maps also reflect the cascade of problems from planned shrinkage and the later targeting by 'Hope VI' and gentrification.

The Bronx, however, shows a much broader collapse, a borough-wide health and safety crisis. It has roots in urban renewal when Herman Badillo was commissioner of the Department of Relocation. He built himself a power base from which to run for higher office by removing large numbers of Puerto Ricans from Manhattan neighborhoods into the South Bronx. Six of the twelve community districts of the Bronx were targeted for this population shift. Badillo, indeed, ran for higher office and was borough president of the Bronx and congressman. In 1969, he ran for the mayoralty and lost by only a little. This near-win drew the eyes of the white political establishment to the Bronx. As a result, the Bronx lost more fire companies proportionately than any other borough during the several rounds of planned shrinkage cuts. See Wallace and Wallace (1998) for details.

Some ZIP Code areas lost well over half their housing in the resulting fire epidemic (for the Bronx, see Fig. 3.3 in Chap. 3). They also lost over half their populations. The refugees moved to the West Bronx and Northwest Bronx. Refugees from the burnout of Harlem moved to the North Bronx and parts of the East Bronx. With a few exceptions, most Bronx ZIP Code areas suffered from the mass migrations and neighborhood social, economic, and political organization and power eroded massively, whether they lost population or suddenly gained large population.

The pattern of COVID death rates across the Bronx may not be real, but an artifact of differential testing of corpses. The presence of a chain of major healthcare facilities across the northern Bronx may have allowed more testing there than in the southern ZIP Code areas, besides offering more opportunities for spread of the virus. What must be considered is the fact that only one Bronx ZIP Code area had a COVID mortality rate below 100/100,000 and only one at 100–150. Even taking into account the fact that the Bronx has the lowest number of ZIP Code areas, we need only look at the maps of the other three boroughs to see that frequency of areas with rates below 150/100,000 reach bottom in the Bronx. This way of considering rates should carry over into the realm of premature death rates. The Bronx had no areas with premature death rates below 100/100,000 and only one with a rate 100–150. The other three boroughs had much higher proportions of their areas with these low rates. Queens enjoyed huge swaths of low-rate areas. Eighteen out of the 35 Manhattan areas had rates below 150. Even Brooklyn had a higher proportion: eight out of 37.

Although a clear pattern of premature mortality concentration in the south-central Bronx emerges from Fig. 2.1b, the entire borough suffers from elevated rates of premature mortality. Figure 2.2b clearly shows that the entire borough suffers from elevated rates of COVID mortality, even though the geographic pattern may underestimate the full affliction because of undertesting of corpses. This very undertesting, which explains the strange associations of COVID case rates with SE factors (opposite direction from the associations in Manhattan) and the lack of association of death rates and SE factors, reflects the lack of responsibility of the authorities to the Bronx. Implementation of planned shrinkage in the other boroughs proceeded more delicately than in the Bronx because of the presence of powerful enclaves. As has been stated above, none of the other three central boroughs was as stripped of fire protection as was the Bronx. The ability of community allostasis to effect homeostasis appears to have been utterly destroyed in the Bronx.

The stripping in other boroughs occurred with surgical slicing and the neighborhoods in Manhattan, Brooklyn, and Queens that lost fire companies tend to have higher premature mortality rates than others nearby. Harlem (Manhattan); Flatbush, Brownsville, East New York, Bedford-Stuyvesant (Brooklyn); the Rockaway Peninsula (Queens) lost fire companies, housing, and population during the 1970s. They also endured the 1975–1993 cascade of public health and safety problems consequent on that destabilization. They showed greater vulnerability to the 2007 Great Recession/ Housing Crisis than other areas (Wallace 2011). They incurred elevated rates of COVID deaths (above 250).

The extremely low rate of COVID deaths in ZIP Code area 10026 (<100) probably results from undertesting the corpses. ZIP Code area 10026 has a poverty rate of 26%, a diabetes mortality rate of 34/100,000, and a rather high premature mortality rate of 241/100,000. There are no hospitals within it and none near it. It sits half-way between Mount Sinai Medical Center and Harlem Hospital. We can conclude that premature mortality and COVID mortality rates in the targeted neighborhoods occur as a product of active oppression over decades.

There are, however, areas of elevated COVID mortality that were not actively targeted by planned shrinkage: the two areas on the Coney Island Peninsula and the cluster of areas around LaGuardia Airport. The Coney Island amusement park and boardwalk deteriorated for many years; the surrounding neighborhood also was neglected. Mayor Michael Bloomberg decided to develop the amusement park. In his campaign to do so, he bid $105 million in tax-payer money to a developer who owned a piece of land that was deemed essential for the revival of the amusement park (Sederstrom 2009). No funding was given to the rest of the peninsula. Housing conditions, schools, healthcare, sanitation, and all the other essentials for a thriving community simply continued their downhill slide. In the eyes of the Mayor, the rest of the peninsula did not exist.

Five of the seven non-Rockaway ZIP Code areas in Queens with COVID mortality rates above 300/100,000 cluster around LaGuardia Airport. Their foreign-born populations represent 54–66% of the area populations, much higher than the borough's unweighted average (44%) and median (42%). Their premature mortality rates were in the 100–150 range and near borough average and median. Their diabetes mortality rates ranged up into the >30 range but most below 20. These areas popped up with high COVID case incidence early in the municipal epidemic and fueled conversations about why.

We have reason to believe that neither NYS DoH nor NYC DoH established long-term relationships with the constituent communities in these areas, and neither made well-planned preparations for disaster that included both the input from the constituent communities and their special needs in terms of communication, encouragement, and urgency. The total population of the five areas comes to about 377,000, well over a tenth of the whole borough. LaGuardia Airport links New York City with other major metropolitan regions and supplies jobs to the surrounding areas. These five areas could have received cases from travelers and, in turn, contributed to the national spread of COVID by infecting travelers.

They fell through the public health and disaster-preparedness cracks. Like the Chasids of Borough Park and the Italians of Bensonhurst, many immigrant populations distrust the authorities and require much attention over long times for a relationship to gain traction. Under normal conditions, gross neglect of these populations allows them to form their own benevolent societies and social clubs, their own PTAs, their own civic associations. The State and the City have a responsibility to include these populations in emergency and disaster planning; they failed. The cluster of elevated COVID mortality rates (and case rates and percents positive) substantiate the failure, neglect, and negligence.

The geography of COVID mortality and of premature mortality reveals massive active and virulent oppression (the Bronx), acutely targeted oppression (poor neighborhoods of color in Manhattan, Brooklyn, and Queens), large-scale neglect and negligence (the immigrant areas of Queens) and small-scale ignoring (Coney Island). This form of corruption, obviously, has cost many thousands of lives.

In the old days of Tammany Hall and Boss Tweed, corruption typically included 'buying votes' by sending coal and food to poor people so that they could survive the winter (Mandelbaum 1990; MacKaye 1940). We are very far from that form of corruption, although catering public policies and public moneys for the very wealthy has remained with us from that era. The very relationship between wealth and power has reverted to that in The Gilded Age, the age of Robber Barons, the age of grinding down labor unprotected by any laws or labor unions, the age of exploitation of tenants and low-income holders of mortgages, and the age of vacuuming away consumers' money, health and safety.

The geographies of COVID death rates and premature death rates that we describe here reflect the differential concentration of oppressive, neglecting, and negligent corruption at all levels of government: federal, state, municipal, and borough. Even the city and state departments of health view these geographies of death as natural, as what should be expected, given the race/ethnicity, low income and wealth, low educational attainment, and lack of political organization and power of the areas. Even the practice of public health is tainted with corruption now, a remarkable and lamentable erosion from the practices of such giants as Griscom (1844), Dubos and Dubos (1952), and Hinkle and Loring (1977).

The New York Academy of Medicine convened a gathering in April 2008 to present data on public health impacts of serial forced displacement. Papers covered topics such as redlining, urban renewal, planned shrinkage, Hope VI, and mass mortgage foreclosures. Several papers appeared in print in the June 2011 issue of *Journal of Urban Health*, the Academy's publication. All these public policies plus gentrification shred communities and destroy public health over several generations. The Bronx's abject SE and public health condition serves as an example of where these policies lead.

However, the boroughs that received 'surgical' targeting may not end up in good condition in the long term (several generations). Manhattan had an extremely tight system of SE factors and public health measures. Brooklyn's system showed somewhat looser ties. Queens which had suffered least from the public policies also had a much looser SE and public health system. Before COVID-19 swept over New York City, Manhattan and Brooklyn strutted their high incomes, high property values, and chic neighborhoods. The phrase 'high on the hog' comes to mind. But they may be yet a couple more examples of Prospero's Castle in more ways than future COVID-19 waves.

Tightly connected systems have no elasticity. They are rigid and brittle. The great ecosystems scientist Holling (1973) introduced the concept of loose vs. tight connections in ecosystems. Healthy ecosystems include high biodiversity and redundancy of ecological niche occupation. They swallow impacts without changing structure or function. This ability is ecological resilience, as opposed to engineering resilience which depends on recovering after an impact. Ecosystems that have been subjected to many impacts or to a few major impacts lose diversity and have tight ties among the remaining species.

They are fragile and in peril of regime change in the face of further impact. The steamrollers of recurrent community destabilization through housing destruction and the real estate economics that fed on the resulting housing famine drained Manhattan and Brooklyn of its human diversity and of the diversity of power bases. Even if a low-income population of color could remain, it had lost its social, political, and economic organization and power. The SE and public health tightness of these two prosperous boroughs reflects this depauperate power structure. They are due for a precipitous fall eventually. The COVID-19 social, economic, and political consequences may trigger this avalanche or any other event or process that tips over this house of cards.

The Bronx exemplifies the outcome of regime change like a scummy green lake that has been eutrophied from untreated sewage and loads of fertilizer runoff. Manhattan with its relatively small population and high population density may be closer to regime change than Brooklyn and much closer than Queens. A warning sign is the level of rent stress. Although Manhattan has the largest number of areas with individual median incomes over the NYC average and median, it has only three areas where less than 30% of the renting households spend more than 35% of their income on rent. In other words, a critical mass of Manhattan renting households are stretched economically even though they may pull in relatively high incomes. Most of the other households have cooperative apartments and struggle with mortgage payments as well as monthly maintenance charges. Probably the rental statistics reflect the whole housing economics picture, including cooperative units. The ruins of Rome's overstretched empire come to mind.

At this moment (early July 2020), the social and political tide that denies the naturalness of high rates of death among people of color and poor and working-class people has risen. It is even reflected in primary election losses of incumbents to those with a vision of reversing the opposing tide of massive corruption. The combination of the horror of COVID illness and death and of the murders of people of color by the State has, for the moment, drawn a sector of white, middle-class groups into the movement that denies the naturalness (Sparks 2020). The US Supreme Court has made several decisions lately that bolster voter suppression, especially suppression of minority, elderly, and working-class voting (example: Nichols 2018). The course of this clash between the powerful few and the disenfranchised many will determine geographies of death in the foreseeable future. Public health researchers and practitioners have a role to play in this clash, if they would acknowledge their responsibility and risk their funding and jobs. In this time of guns, they may also risk their lives.

References

CDC (2020) https://data.cdc.gov/nchs/provisional-COVID-19-death-counts-by-sex-age

Dubos R, Dubos J (1952) The white plague: tuberculosis, man and society. Little, Brown, and Company, Boston

Friedman M (2019) https://www.nytimes.com/2019/04/23/opinion/my-fellow-hasidic-jews-are-making-a-terrible-mistake-about-vaccinations

Geronimus A (1996) Black/white differences in the relationship of maternal age to birthweight: a population-based test of the weathering hypothesis. Soc Sci Med 42:589–597

Griscom J (1844) The sanitary condition of the laboring population of New York. Arno Press, New York. Reprinted 1970

Hasidic Funeral (2020) https://www.nytimes.com/2020/04/28/nyregion/hasidic-funeral-coronavirus-de-blasio.html

Hinkle L, Loring W (1977) The effects of the man-made environment on health and behavior. Dhew. Publication No. (CDC)77-8318. Center for Disease Control, Washington DC

Holling CS (1973) Resilience and stability of ecosystems. Ann. Rev. Ecol. Syst. 4:1–23

Kao K, Sabin M (2016) Type 2 diabetes in children and adolescents. Aust Fam Phys 45:401–406

Kim G, Jee S, Pitkart H (2018) Role of allostatic load and health behaviors in explaining socioeconomic disparities in mortality: a structural equation modeling approach. J Epidemiol Community Health 72:545–551

MacKaye M (1940) Four gentlemen of Manhattan. In: Churchill A (ed) A treasury of modern humor. Tudor Publishing Company, New York, pp 91–113

Mandelbaum S (1990) Chapter 5. Boss Tweed's New York. Elephant Paperbacks, Chicago

McEwen B (2015) Biomarkers for assessing population and individual health and disease related to stress and adaptation. Metabolism 64:S2–S10

Mitchell D (2017) Growth in patients with type 1 diabetes. Curr Opin Endocrinol Diabetes Obes 24:67–72

Nichols J (2018) https://www.thenation.com/article/archive/supreme-court-give-green-light-massive-voter-suppression

Nursinghome411 (2020) https://nursinghome411.org/ny-nursinghome-COVID-data

NYC DoH (2020) https://www1.nyc.gov/site/COVID-19-data.page

Romeo R, McEwen B (2006) The neonatal and pubertal ontogeny of the stress response: implications for adult physiology and behavior. Chapter 8. In: Arnetz B, Ekman R (eds) Stress in health and disease. Wiley-VCH, Weinheim

Sederstrom J (2009) https://www.dailynews.com/new-york-brooklyn-mayor-bloomberg-final-bid-coney-island-property

Sparks G (2020) https://www.cnn.com/2020/06/18/politics/protests-polling-support-movement-policies-kaiser-quinnipiac.html

Tan M, Mamun A, Kitzman H, Mandapati S, Dodgen L (2017) Neighborhood disadvantage and allostatic load in African American women at risk for obesity-related disease. Prev Chron Dis 14:E119. https://doi.org/10.5888/pcd14.170143

Wallace D (1994) The resurgence of tuberculosis in New York City: a mixed hierarchical and spatially diffused epidemic. Am J Public Health 84:1000–1002

Wallace D (2011) Discriminatory mass de-housing and low-weight births: scales of geography, time, and level. J Urban Health 88:454–468

Wallace D, Wallace R (1998) A plague on your houses: how New York City was burned down and national public health crumbled. Verso, New York and London

Yeshiva Investigation (2019) https://www.nytimes.com/2019/12/25/opinion/yeshivas-investigation-de-blasio.html

Chapter 3
Prospero's New Castles: COVID Infection and Premature Mortality in the NY Metro Region

3.1 Introduction

The explosion of the virus SARS-CoV-2 from feral Chinese bat populations into pandemic COVID-19 is simply the latest case of a much larger public health dynamic driven by longstanding colonial, neocolonial, neoliberal and agribusiness policies and practices (e.g., Wallace 2016; Wallace et al. 2020). Under such conditions, pulses of novel pathogens circle like vultures waiting to land on the gangrenous limbs of fallen and emerging empires. Appearance of some analog to African Swine Fever, or Nipah virus, with fatality rates of 50% or more, appears simply a matter of time, under current power structures.

Emergence of HIV/AIDS from African SIV-infected primates into human populations provides a well-studied example of a 'slow plague' (Gould 1993) whose spread dynamic was studied via models of diffusion in a commuting field (Gould and Wallace 1994; Wallace et al. 1997, 1999) can easily be extended to other disease systems enmeshed in the daily journey-to-work. Here, we examine the pattern of COVID-19 deaths through May, 2020 in the New York Metropolitan Region, a 24-county system characterized in Figure 2 of Gould and Wallace (1994) and listed in the Appendix Data Table.

The context for this analysis is the full 'AIDS Syndemic' of 1980–1993 that saw AIDS, tuberculosis (TB), violent crime, and low birthweight regionalized across the New York Metropolitan Region (NYMR), largely as a consequence of 'planned shrinkage' policies aimed at the dispersal of minority community voting blocs in New York City during the 1970s (Wallace 1988; Wallace and Wallace 1998, 1990; Wallace 1990).

The original version of this chapter was revised: The incorrect Fig. 3.1 has been corrected now. The correction to this chapter is available at https://doi.org/10.1007/978-3-030-59624-8_6

3.2   The AIDS Syndemic

The central insight necessary for understanding social and geographic diffusion in a metropolitan region is recognition that the journey-to-work is a powerful environmental index of the underlying processes. The US Census records, at the county level, the number of those working in and commuting to work between, individual counties. For the NYMR, this permits periodic construction of a 24 by 24 'commuting matrix' of the observed numbers. Such a matrix, however, is too large to be easily used. The central trick is to first normalize the matrix to unit row sum, generating a stochastic matrix $\mathbf{P}$ representing a Markov process, and then to find the 24-element equilibrium distribution of that matrix, i.e., the left eigenvector corresponding to the eigenvalue 1. This can be done by solving the relation $\boldsymbol{\mu} \cdot \mathbf{P} = \boldsymbol{\mu}$ for $\boldsymbol{\mu}$. However, 'it is not difficult to show' that this is equivalent to any row of the power matrix $\lim_{k \to \infty} \mathbf{P}^k$ (Kemeny and Snell 1976). Gould and Wallace (1994) and Wallace et al. (1997) use the equilibrium distribution calculated from 1980 US Census data, found to be essentially similar to that derived from the 1990 Census.

Wallace et al. (1997) examine the 1990-era regional 'synergism of plagues' (Wallace 1988) afflicting the NYMR. In that work, the dependent variates were the logs of the population rates of violent crime 1985, AIDS through 1990, tuberculosis 1985–1992, and rate of low-weight births for 1988. The independent variates were the logs of the 1980 equilibrium distribution per unit area for the counties, and the county rate of poverty for 1980. In that paper, a multivariate analysis of covariance (MANCOVA) for violent crime, AIDS and tuberculosis found a single expression in μ / A and the percent living in poverty accounted of over 90% of model variance across the full 24-county NYMR. The rate of low-weight births fits a somewhat different pattern, but still fit a similar model at over 90% of variance. In all cases, Manhattan represents the upper-right most data point of the relations, indicating its centrality in the spread dynamic.

The work of Wallace et al. (1997) makes explicit the 1990-era synergism of plagues (Wallace 1988)—now called a 'syndemic' (e.g., Singer et al. 2017)—across the NYMR. It represents a time-smoothed mixmaster regionalization of pathologies emerging from profound disturbance of the apex conurbation of the American Empire, the vast policy-driven 'planned shrinkage' devastation of minority communities in four of the five New York City counties that took place in the 1970s (Wallace and Wallace 1998; Wallace et al. 1999).

The magnitude of this disturbance is indexed by Fig. 3.1, showing the percentage loss of occupied housing units in the Bronx section of New York City to contagious fire and abandonment between 1970 and 1980. The map is of 'health areas', aggregates of census tracts by which health data are reported. The South-Central region—a minority voting bloc—lost from 36 to as much as 81% of occupied housing in that time. Large minority sections of Brooklyn, Queens, and Manhattan suffered analogous housing loss, dispersing and fragmenting all forms of organized community activity, including the ability to socialize adolescents into the adult world of work, which, in large measure, ceased to be available within the affected communities.

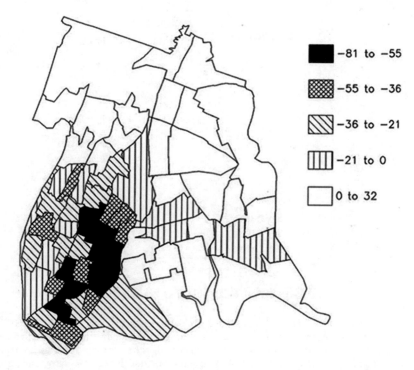

Fig. 3.1 Percent loss of occupied housing units in the Bronx 1970–1980. Other minority voting blocs in Manhattan, Brooklyn, and Queens were similarly devastated by the politically-targeted withdrawal of housing-related municipal services, especially fully-staffed fire companies (Wallace and Wallace 1998)

This perturbation was of sufficient magnitude to drive violent crime and low birthweight—most often indices of ongoing chronic conditions—into contagious propagation throughout the New York Metropolitan Area.

Wallace et al. (1999) examine the impact of this profound disturbance of the apex of the US urban hierarchy on the spread dynamics of AIDS across the 25-largest US metropolitan regions, driven on a longer time scale by a probability of contact vector indexed by migration rates..

COVID-19 comes, a half-century later, into this profoundly wounded sociogeographic landscape.

3.3 The COVID-19 Pandemic

Using the data of the appendix, Fig. 3.2 conducts an analysis based on that of Wallace et al. (1997). The dependent variate is the log of the COVID-19 death rate, while, again, the central independent variates are the logs of the equilibrium

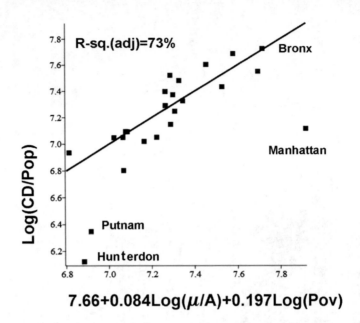

$$7.66+0.084Log(\mu/A)+0.197Log(Pov)$$

Fig. 3.2 From the data of the chapter appendix. The dependent variate is the log of the COVID-19 death rate. The independent variates are the logs of the equilibrium distribution per unit area and of the percent living in poverty for the NYMR, updated. It is necessary to break the system into two quite distinct parts. Putnam, NY; Manhattan; and Hunterdon, NJ counties have fallen from of the regionalization. The line shows the regression fit to the remaining 21 counties, accounting for 73% of the adjusted variance. Bronx county now represents the upper-most data point

distribution per unit area and of the percent living in poverty for the 24 counties of the NYMR used in Wallace et al. (1997), updated as indicated. Now, however, it proves necessary to break the system into two quite distinct parts, in distinct contrast to the pattern of AIDS, TB, violent crime and low birthweight found in Wallace et al. (1997) in which Manhattan was the dominant data point within a fully unified system.

That is, in contrast to the earlier results, Putnam, New York; New York (i.e., Manhattan); and Hunterdon, New Jersey counties have dropped out of the regionalization for the spread of COVID-19. The straight line is the fit to the remaining 21 counties, accounting for some 73% of the (adjusted) variance. Here, Bronx county now represents the upper-most data point.

It is interesting to note that, for the COVID-19 analysis, as for the much earlier AIDS syndemic study, indices of ethnicity/race at the county level did not provide greater explanatory power.

In contrast to the results of Wallace et al. (1997) however, examination of chronic, life-trajectory pathology, shows a markedly different pattern. Figure 3.3 relates premature mortality—the death rate per 100,000 before the age of 65—to the same variates as in Fig. 3.2, i.e., μ/A and the percent living in poverty. Here, for largely chronic conditions, the regression coefficient for μ/A has a *negative* sign.

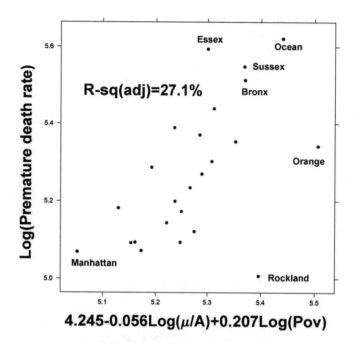

4.245-0.056Log(μ/A)+0.207Log(Pov)

Fig. 3.3 From the data of the chapter appendix. The dependent variate is now the log of the rate per 100,000 of deaths under the age of 65, averaged for 2014–2016. The independent variates are as above. For this regression model, however, the 'outliers' have not been removed. On the whole, life course trajectory is markedly improved by access to the economic engine of the central city. The anomalous position of the Bronx in this relation suggests, however, the continued operation of powerful strategies and traditions of discrimination, in the context of the persistent 'avalanche' effects of the policies of 'planned shrinkage' leading to the disaster of Fig. 3.1

Thirty years after the results of Wallace et al. (1997), in which Manhattan strongly dominated overall diffusion of multiple pathologies throughout the New York Metropolitan Region, we see Manhattan at the lower left, not the upper right. By 2015, life course trajectory in the New York Metropolitan Region has been *improved* by access to the economic engine of the central city, while, in contrast, for rates of COVID-19 infection, as earlier for AIDS and tuberculosis, risk was, on the whole, increased by the daily mixmaster of that economic engine. The anomalous position of the Bronx toward the upper right of both figures, however, suggests the influence of powerful policies of marginalization and discrimination within New York City itself.

Consonant with the 'disaster avalanche' analysis of Wallace and Wallace (2016, Ch. 5), we see something of the long-term impacts of the 'planned shrinkage' policies aimed at causing the devastated landscape of Fig. 3.1, the hollowed-out minority voting blocs of the Bronx. That county's position at the upper right of both Figs. 3.2 and 3.3—nearly 50 years later—constitutes an index case history. Those figures relate—and contrast—the early stage of an emerging infection and the influence of long-term chronic pathologies to poverty and position in the commuting field. The effects of the deliberate destruction of Fig. 3.1 will persist until real, long-term, relief efforts become manifest. The mechanism for such persistence can be described in terms of an ecosystem 'resilience shift' (Holling 1973), analogous to the eutrophication of a pristine lake by agricultural or sewage runoff.

In sum, institutions, commercial enterprises, their topologically rich networks and communities, and their larger, embedding social structures, must engage in cognitive process to address rapidly changing patterns of challenge and opportunity, and, more slowly, incorporate the learning necessary for successful adaptation to shifts in larger scale evolutionary selection pressures. Failure of cognition, or of learning/adaptation, can be triggered by environmental challenges that drive such structures into highly persistent ground states where cognitive or learning/adaptational process becomes pathologically fixated, initiating a developmental pathway to failure.

3.4   Diffusion in a Commuting Field

Following Gould and Wallace (1994) and Wallace et al. (1997), we explore diffusion between and among geographic regions embedded within a commuting field, using a simplified version of the previous model in which regions are assumed to have been appropriately normalized by the underlying poverty index representing social hierarchy.

For each region let $\Delta I(t)$ be the number of infected at time t, take ΔN as the population, $\Delta \mu$ the equilibrium Markov distribution, ΔA the area, and $b(t)$ be an appropriate fitting parameter. Then the poverty-normalized spread dynamic is described by the empirical relation

$$\log\left(\frac{\Delta I(t)}{\Delta N}\right) = m\log\left(\frac{\Delta \mu}{\Delta A}\right) + b(t) \tag{3.1}$$

Dimensional considerations, however, require imposition of normalizing terms, reexpressing t as $\tau \equiv t/T_0$, so that

$$\log\left(\frac{\Delta I(\tau)}{\Delta N}\right) = m\log\left[\left(\frac{\Delta \mu}{\Delta A}\right)A_0(\tau)\right] + h(\tau) \tag{3.2}$$

where T_0 is the relaxation time characterizing the spread process and $A_0(\tau)$ is now the characteristic area of the process which will increase with time.

Rearranging terms,

$$\Delta I(t/T_0) = \Delta N\left[\left(\frac{\Delta \mu}{\Delta A}A_0(t/T_0)\right)\right]^m \exp[h(t/T_0)] \tag{3.3}$$

The meaning of this relation is clarified by comparing it with the simplest two-dimensional diffusion process.

Let $i \approx \Delta I / \Delta A$ be the area density of individuals at time t at radius r from a central point where ΔN mobile and reproducing individuals are placed at time 0. Near the beginning, before limits are approached, the simplest symmetric two-dimensional population growth relation is

$$\partial i / \partial t = D \nabla^2 i + \alpha i \qquad (3.4)$$

where D is the diffusion coefficient and α the population's growth rate.

This relation has the radially symmetric solution

$$i(r, t) = \frac{\Delta N}{4\pi Dt} \exp\left(\alpha t - \frac{r^2}{4Dt}\right) \qquad (3.5)$$

rewritten for a small ΔA located at the radial distance r from the center as

$$\Delta I(r, t) \approx \Delta N \frac{\Delta A}{4\pi Dt} \exp\left(\alpha t - \frac{r^2}{4Dt}\right) \qquad (3.6)$$

This dimensionless form of the previous relation is analogous to the dimensionless empirical result if $\alpha = 1/T_0$ and $Dt = A_0$.

Dt is a variance measure with the dimension of an area, representing the enlargement of the characteristic area of the population as it diffuses and reproduces.

As infection proceeds in a commuting field, the characteristic area A_0 increases from a block to a zip code, to a community district, to a county, to a metropolitan region. Eventually, by other mechanisms, the characteristic area emerges across larger entities composed of linked metropolitan regions. This is a level of organization at which diffusion cannot proceed via the daily journey-to-work and will, if again AIDS provides guidance, likely need a different approach (Wallace et al. 1999).

3.5 Toward a National Model

Wallace et al. (1999) explore diffusion of the AIDS 'slow plague' across the 25 largest US metropolitan regions. The driving mechanism was, however, not as described above, i.e., a mixmaster effect similar to the daily commute dominating metropolitan regional dynamics (Wallace et al. 1997). National scales of time, space, and population, are not those of the counties within a single metropolitan area, and consequently reflect different underlying dynamics.

It was, however, still possible to generate a 25×25 stochastic matrix, $\mathbf{P}$ as above, based on census data describing, not the daily journey-to-work, but migration within and between counties for 1985–1990, as aggregated into metro regions. Exploratory data analysis quickly demonstrated that it was not the equilibrium distribution of the associated Markov chain, essentially found by raising the matrix to a high power, that dominated disease spread. It was a single vector representing the probability of contact with the largest US metro region, the NYMR, driving the diffusion of disease between metropolitan regions. That is, the 25-element vector $\mathbf{P}[1;\ ]$ rather than μ satisfying $\mu \cdot \mathbf{P} = \mu$, drove the national diffusion of AIDS, as modulated by local indices of susceptibility. Those, too, differed from the NYMR AIDS analysis.

While the probability defined by the migration-from-NYMR vector served as the 'global' measure, again exploratory data analysis found two local indices of susceptibility. These were the violent crimes for 1991, and the number of manufacturing jobs for 1987 divided by that for 1972. It was, indeed, possible to construct not only regression models, but also to carry out a MANCOVA analysis for the diffusion of AIDS across the United States. That is, a single index constructed from these variates served to characterize the number of AIDS cases through April 1991, and between April 1991 and June 1995, for the 25 largest US metropolitan areas. The rising case load in the NYMR, the powerful apex of the US urban hierarchy, literally pulled up, in parallel, the case loads for 24 of the smaller, but still closely-linked, metro regions.

Wallace et al. (1999) explored a MANCOVA for the 25 largest US metro regions of the log of the AIDS cases vs. the index X

$$X = 0.764 \log(USVC91) + 0.827 \log(USME87/USME72) + 0.299 \log(\mathbf{P}[1;\ ]) \tag{3.7}$$

USVC91 was the number of reported violent crimes for 1991, USMExx the number of metro regional industrial jobs for the year 19xx, and $\mathbf{P}[1;\ ]$ the probability of contact with the New York Metro Region calculated from migration data 1985–1990.

Wallace et al. (1999) found that, for the AIDS pandemic, as went the apex of the US urban hierarchy, so, ultimately, went the nation. The implication is that emerging infection first appearing in any US metropolitan area will be rapidly drawn into the apex of the associated urban hierarchy, incubated in marginalized populations, and then blown back down along that hierarchy. Understanding the diffusion dynamics of the full COVID-19 disaster—including the inevitable 'avalanche' effects of COVID-19 on other causes of mortality—in the United States will require acquisition of reliable county-level data on deaths in excess of those expected from historical patterns, properly aggregated, and using updated indices of regional contact dynamics. . We are, of course, as of this writing, still in the early stages of the COVID-19 pandemic, and application of these methods may not provide as good a fit as found in the Wallace et al. (1997) AIDS syndemic regionalization study. Indeed, metropolitan regional examples of Prospero's Castle should be expected as outliers at this early date. However, as has been said in another context, a rising tide lifts all ships.

A 'sloshing' dynamic is perhaps to be expected, the first phase involving aspiration into the largest metropolitan areas from outlying hot spots, followed by blowback down the US urban hierarchy, and then subsequent reseeding of the largest metro areas back up the hierarchy until an effective vaccine is widely adopted.

3.6 Discussion

What has happened in Fig. 3.2? How has Manhattan become part of an apparent urban/suburban disjunction for the NYMR with respect to the COVID-19 death rate? Can this disjunction last as COVID-19 enters more fully into regional sociogeographic dynamics over the next few years?

Although the healthcare facilities of Upper Manhattan whose catchment area includes the poor of Harlem, Washington Heights, and the Bronx, were greatly burdened by the April to May 2020 wave of COVID-19 cases—indeed, with senior staff driven to suicide—and while Manhattan's poverty-stricken Chinatown was similarly afflicted, Manhattan as a whole seemed curiously removed. What has changed between 1980 and 2020?

There is an all too simple index. In 1980, a middle-class 3-bedroom coop apartment just north of Columbia University at 116th Street. sold for well under $10,000. The same apartments, in buildings 40 years older, now sell for well over $1,000,000. Putnam and Hunterdon Counties are, in their own ways, similarly notorious.

The 2020 median incomes of Manhattan, Putnam, and Hunterdon counties are, respectively, $66,739, $99,608, and $110,969. For most of Manhattan south of 110th Street, the median income is well over $100,000, the Lower East Side/Chinatown excepted. By comparison, that of the Bronx is $38,467.

Table 2.6 in Chap. 2 compares COVID-19 indices, as of this writing, for the 24 Bronx and 35 Manhattan ZIP codes with more than 10,000 residents. The two Boroughs have, respectively, populations of 1.42 and 1.63 million, and poverty rates of 30.7 and 17.9%.

Figure 3.2 represents, in a sense, the early days of pandemic spread. The work of Wallace et al. (1997), by contrast—with Manhattan as the peak driver—caught AIDS and tuberculosis as long-entrenched infections, well-integrated into the spatial ecology of the NYMR.

Prospero's castle, the affluent regions of Manhattan, together with Putnam and Hunterdon counties, is, at the date of this writing, locked down and telecommuting during only the very first stages of the COVID-19 pandemic. Experience of the last century, from the second wave of the 'Spanish flu'—better described as Kansas hog influenza—through the many subsequent influenza outbreaks, suggests that, over time, the walls of that castle will indeed by breached, that the dynamic mixmaster indexed by the commuting field links every geographic entity of the New York Metropolitan Region with every other at much less than six degrees of separation. 'Telecommuting' by the declining number of affluent US workers can only slow, but not halt, that dynamic process.

In plain words, the outliers of Fig. 3.2 will inevitably be drawn into the general system of the other 21 counties of the NYMA by powerful mechanisms of sociogeographic diffusion.

Similar dynamics must, in fact, ultimately characterize an emerging pathogen across the full system of metropolitan regions, first drawn into the apex of the urban hierarchy, and then blown back and forth along it. Concentration is not containment, but the central mechanism for general spread.

3.7 Appendix: The Data Set

Pop	A	Mu	Pov	CD	County	PDR
1.629	22	0.551372	17.9	2008	New York, NY	159.1
1.418	42	0.027696	30.7	3211	Bronx	247.9
2.560	70	0.069368	23.2	4872	Kings	194.5
2.253	109	0.074503	15.1	4907	Queens	171.4
0.476	59	0.003974	12.5	724	Richmond	230.2
0.967	438	0.035023	9.6	1359	Westchester	163.1
0.326	175	0.004757	14.6	519	Rockland	149.5
0.098	231	0.000712	5.3	56	Putnam	200.9
0.385	826	0.001973	12.8	444	Orange	209.0
1.357	287	0.037825	6.2	1991	Nassau	178.1
1.477	911	0.028168	7.0	1654	Suffolk	219.1
0.917	632	0.036405	10.0	1167	Fairfield, CT	167.7
0.937	237	0.024915	6.8	1529	Bergen, NJ	163.1
0.676	46	0.019291	14.3	1143	Hudson	181.3
0.558	103	0.010351	7.8	1030	Union	197.8
0.800	127	0.020277	14.9	1605	Essex	268.9
0.503	187	0.008410	13.6	892	Passaic	211.4
0.494	470	0.014572	4.7	595	Morris	162.8
0.331	305	0.007820	4.9	399	Somerset	159.6
0.830	316	0.013621	8.2	959	Middlesex	176.5
0.621	472	0.005207	6.7	559	Monmouth	215.1
0.601	641	0.001764	9.7	692	Ocean	276.3
0.125	426	0.001381	4.4	57	Hunderdon	187.7
0.141	526	0.000514	5.2	145	Sussex	256.8

Population in millions; county area in square miles; equilibrium distribution using 2015 US Census journey-to-work data; percent of population living in poverty 2015; reported COVID-19 deaths June 1, 2020; county name; and premature death rate (<65), averaged for 2014–2016. There is an important caveat: the reported numbers of COVID-19 deaths are only as reliable as the confirmation processes. They may be underestimates, with the degree of underestimation related to the poverty rate and population of color of the county.

References

Gould P (1993) The slow plague: a geography of the AIDS pandemic. Blackwell, London

Gould P, Wallace R (1994) Spatial structures and scientific paradoxes in the AIDS pandemic. Geogr Ann B 76:105–116

Holling C (1973) Resilience and stability of ecological systems, Ann Rev Ecol Syst 4:1–23

Kemeny J, Snell J (1976) Finite Markov chains. Springer, New York

Singer M, Bulled N, Ostrach B, Mendenhall E (2017) Syndemics and the biosocial conception of health. Lancet 389:941–950

Wallace D. Wallace R (1998) A plague on your houses. Verso, New York

Wallace R (1988) A synergism of plagues: 'planned shrinkage', contagious housing destruction and AIDS in the Bronx. Environ. Res. 47:1–33

Wallace R, Wallace D (1990) Origins of public health collapse in New York City: the dynamics of planned shrinkage, contagious urban decay and social disintegration. Bull. NY Acad Med 66:291–343

Wallace R (1990) Urban desertification, public health and public order: planned shrinkage, violent death, substance abuse and AIDS in the Bronx. Soc Sci Med 31:801–813

Wallace R, Wallace D, Andrews H (1997) AIDS, tuberculosis, violent crime and low birthweight in eight US metropolitan areas: public policy, stochastic resonance and the regional diffusion of inner-city markers. Environ Plan A 29:525–555

Wallace R., Wallace D, Ullmann JE, Andrews H (1999) Deindustrialization, inner-city decay, and the hierarchical diffusion of AIDS in the USA: how neoliberal and cold war polices magnified the ecological niche for emerging infections and created a national security crisis. Environ Plan A 31:113–139

Wallace RG, Wallace R (2016) Neoliberal Ebola: modeling disease emergence from finance to forest and farm. Springer, New York

Wallace RG (2016) Big farms make big flu: dispatches on infectious disease, agribusiness, and the nature of science. Monthly Review Press, New York

Wallace R (2010) Cognitive dynamics on Clausewitz landscapes: the control and directed evolution of organized conflict. Springer, New York

Wallace R, Liebman A, Bergmann L, Wallace RG (2020) Agribusiness vs. public health: disease control in resource-asymmetric conflict. https://hal.archives-ouvertes.fr/hal-02513883

Chapter 4
Pandemic Firefighting vs. Pandemic Fire Prevention

4.1 Introduction

Firefighting is, beyond its utter necessity, both heroic and romantic: while others flee the burning building, firefighters are, more or less calmly and most definitely, with professional care and discipline, going inside. They often comment on the surreal nature of this disjunction.

But, if firefighters are given sufficient resources, under normal conditions, most fires can be contained with minimal casualties and property destruction. However, that containment is critically dependent on a far less romantic, but no less heroic enterprise, the persistent, ongoing, regulatory efforts that limit building hazard through code development and enforcement, and that also ensure firefighting, sanitation, and building preservation resources are supplied to all at needed levels.

If industry lobbyists are permitted to weaken building codes, for example allowing the use of aluminum wiring and polyvinyl chloride (PVC) electrical insulation and construction materials in places of business or residences (Wallace 1990), if 'demographic engineers' are allowed to cut firefighting resources in communities of color with the intent of dispersing voting blocs (Wallace and Wallace 1998), then there are deadly consequences. Under such conditions, individual fires very often cannot be contained, and indeed, can create 'South Bronx' fire/abandonment epidemics that consume vast stretches of urban landscape, triggering massive loss of life, both directly, and by long-term 'avalanche' social and public health mechanisms (again, Wallace and Wallace 1998).

In sum, code development-and-enforcement that prevents fires is far preferable to actual firefighting, necessary as that may always be.

A recent influential preprint by Neil Ferguson's group at the United Kingdom's Imperial College (Ferguson et al. 2020) provides a singular example of the labor and capital-intensive 'firefighting' necessary to control pandemics:

D. Wallace, R. Wallace, *COVID-19 in New York City*, SpringerBriefs in Public Health, https://doi.org/10.1007/978-3-030-59624-8_4

The global impact of COVID-19 has been profound, and the public health threat it represents is the most serious seen in a respiratory virus since the 1918 H1N1 influenza pandemic. Here we present the results of epidemiological modeling which has informed policymaking in the UK and other countries in recent weeks. In the absence of a COVID-19 vaccine, we assess the potential role of a number of public health measures – so-called non-pharmaceutical interventions (NPIs) – aimed at reducing contact rates in the population and thereby reducing transmission of the virus. In the results presented here, we apply a previously published microsimulation model to two countries: the UK (Great Britain specifically) and the US. We conclude that the effectiveness of any one intervention in isolation is likely to be limited, requiring multiple interventions to be combined to have a substantial impact on transmission...

...and so on, absent discussion of the contextual factors producing the epidemic in the first place that, in all fairness, are not yet well known, but seem to implicate land-use interpenetrations: Wu et al. (2020) and Zhou et al. (2020) suggest pathogen escape from low-level endemic infection in bat populations.

While Shen et al. (2020) critique the details of Ferguson et al.—concluding that "...their model is several degrees of abstraction away from what is warranted by the situation"—there is a deadly serious additional problem. COVID-19, while the first high fatality rate infection surge to reach deadly pandemic status in the last century, is not the first such surge to try, and will most certainly not be the last to succeed.

4.2   Pandemics in Waiting: Neoliberal Land Use Policies

Ferguson et al., by necessity, were focusing on the United Kingdom's problem of controlling the latest rapidly evolving infectious disease disaster. Having to control wave after wave of such outbreaks means not having sufficient resources or attention placed on preventing pandemics. This is analogous to the US Pentagon and the Soviet Union's military parasitizing the pool of scientists and engineers needed to ensure the stability and productivity of domestic industrial production (e.g., Wallace et al. 1999 and references therein).

Taking the larger perspective, Wallace and Wallace (2015) concluded a study of agriculturally-driven pathogen evolution and spread—exploitative land use policies synergistic with factory farms for hogs and fowl—in these terms:

> The blowback from individual and collective decisions on automobile ownership and usage in the UK and the USA appears to have included unleashing polio epidemics among populations that were previously screened from infection by relative isolation. The public health consequences of individual and collective decisions regarding inexpensive animal protein [i.e., factory farms] threaten to be far more serious. Absent fundamental political change... reversing the exploitative relationships instantiating current global policies, the blowback harvests of infection will likely continue to accrue until one of the more virulent of strains now evolving wipes out a good portion of humanity.

Context counts for pandemic onsets, and current political structures that allow multinational agricultural enterprises to privatize profits while externalizing and socializing costs, must become subject to 'code enforcement' that reinternalizes those costs if truly mass-fatal pandemic disease is to be avoided in the near future.

Monoculture, monocropping, animal monogenomics, permit 'economies of scale' for large agricultural corporations that can be used to ensure market dominance—in particular driving small-holders to economic extinction—but at the price of repeated outbreaks of animal pathogens that would otherwise have been smoothed out by variabilities. As Wallace and Wallace (2015) put it,

> By their diversity in time, space, and mode, traditional and conservation agricultures can create barriers limiting pathogen evolution and spread analogous to a sterilizing temperature. Large-scale monocropping and confined animal feeding-lot operations remove such barriers, resulting, above agroecologically specific thresholds, in the development and wide propagation of novel disease strains...
>
> ...[W]hile spatial fragmentation of natural ecosystems can drive wanted species to extinction, proper mosaic design of agricultural systems – fragmentation in time, space, and community structure – can limit the rate of new pathogen evolution and constrain their populations to low endemic levels. In contrast, the inference is that large-scale intensive husbandry, expanding growth and transport of monoculture livestock, is epizootically unsustainable. Increasingly accessed and agroecologically pauperized landscapes should select for pathogens of increasing inadmissibility and virulence.

The last sentence may, in fact, be found to apply to the emergence of COVID-19. Again, see Wu et al. (2020) and Zhou et al. (2020), in particular their remarks on the possibility of a bat host for the virus..

Wallace et al. (2020) provide explicit, and relatively simple, models of pathogen 'stochastic sterilization' arising from variations in time, space, and genetic structure. They summarize this as follows:

> The potential applications in time, space, and genetics are foundational. Other 'stochasticity temperatures' more conducive to disease control can be regionally planned. Crop rotation, agrobiodiversity, and ecological pest control can be incentivized by top-down market regulation and trade/tax structures. The city of Belo Horizonte, a city of 2.5 million in Brazil, implemented such a program... In conjunction with a state extension service, the city's Municipal Under-Secretariat of Food and Nutritional Security helped establish agroecological practices among outlying rural smallholders, and protect the mega-biodiverse Mata Atlantica-Cerrado, by guaranteeing a market and set prices subsidized for low-income urban consumers. Mosaic agroecologies at large spatial and population scales, by their rich environmental stochasticities able to preempt most large-scale diseases before they emerge, appear feasible with appropriate government support.

After some further theoretical development, involving applications of cognitive and control theory to institutional function, Wallace et al. (2020) make recommendations for action:

> Failures in public health's conflict with agribusiness have produced collateral damage in the millions of people and animals hit by deadly pathogens and other environmental spillovers propagating far beyond the farm gate... The clash can be resolved in the broader population's favor, permitting social reproduction generation to generation in the face of what presently are by all accounts a series of protopandemics. To that aim, in the series of models presented here we examined strategies for the 'sterilization' of infectious disease at the population level via stochasticities that conflict with current agribusiness 'economies of scale' privatizing profits and socializing costs. We have used, in addition, the perspectives of control theory and of the asymmetric conflict between entrenched agribusiness interests rich

in resources and public health entities constrained by the many pressures exerted by those very resources on the structures of governance. Asymmetric conflict has been surprisingly successful in military realms. Abduction of appropriate strategies from those examples may also prove successful for the prevention and control of mass-fatal, agribusiness-driven pandemics. Indeed, we argue that the ability to rapidly and systematically adapt strategies and tactics is as central to success in public health as it is in armed conflict.

4.3   Pandemic Penetrance

Following Wallace and Wallace (2016, Ch.5), it is not difficult to make a crude estimate of maximal pandemic penetrance at today's world population. Pandemic influenza of 1918 is thought to have affected some 500 million out of a total population of about 1.5 billion, a 1/3 penetrance (Taubenberger and Morens 2006). That benchmark allows calibration of Kendall's simple epidemic model with removal.

Assuming a total of N individuals in the population of interest, classified as X susceptible, Y infective, and Z removed, the dynamic equations of the Kendall model (Bailey 1975) are

$$dX/dt = -\beta XY$$
$$dY/dt = (\beta X - \gamma)Y$$
$$dZ/dt = \gamma Y$$
$$N = X + Y + Z \tag{4.1}$$

β is the rate of infection and γ the removal rate. Letting $\rho = \gamma/\beta$, no epidemic can spread if the removal rate is greater than the infection rate, i.e., $X(0) < \rho$, where $X(0)$ is the susceptible population at time $t = 0$, and N the total population.

Following Kendall, Bailey (1975, eq. 6.22) shows that, if I is the proportion of the total number of susceptibles that finally contract the disease, assuming a small initial number of infectives and a large N, then

$$N/\rho \equiv s = -\log[1 - I]/I \tag{4.2}$$

This equation has the solution

$$I = 1 + \frac{W[-s\exp(-s)]}{s} \tag{4.3}$$

where $s = N/\rho$ and W is the zero-order branch of the Lambert W-function, which solves the relation $x = W(x)\exp[W(x)]$. Equation (4.3), however, also represents the size of the 'giant component' in a random-network percolation model (e.g.,

Fig. 4.1 The fraction of a susceptible population infected by a contagious process on a random network as a function of the ratio of total susceptible population to the critical population size, N/ρ, from Eq. (4.3). In 1918, some 1/3 of the total population of 1.5 billion became infected, suggesting a critical population of about 1.233 billion. By 2015, the total population has reached about 7 billion, suggesting a ratio N/ρ of 5.68. This leads to a pandemic penetrance of .9965, under 1918 contact probabilities

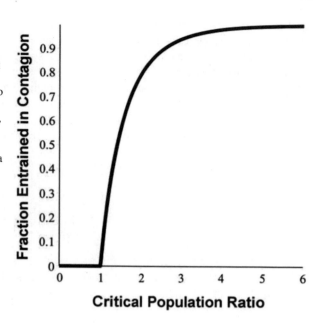

Parshani et al. 2010; Gandolfi 2013). See Fig. 4.1, which shows I as a function of the ratio N/ρ.

A simple calculation can be based on the 1918 observations as a boundary condition. In 1918, some 1/3 of the total population of 1.5 billion became infected, suggesting a critical population ρ of about 1.233 billion, under 1918 travel and other contact probability conditions. By 2020, the total population has reached about 7 billion, suggesting a ratio N/ρ of 5.68. This leads to a pandemic penetrance of about 0.9965, *under 1918 contact probabilities*. Social distancing and contact tracing policies might shave this somewhat, but pandemic penetrance will still likely be substantial for any infection with a long latent-but-infectious period.

Indeed, travel patterns are much tightened since 1918: jets rather than steamers. For highly connected networks, as in the nested fractals of road systems, air travel, and 'strong ties' social networks—stars of stars of stars—there may be no threshold condition whatsoever.

4.4   Refugia: Failure of the 'R_0' Model

A popular deterministic model of epidemic spread focuses on 'R_0', the rate at which an incipient infection grows in time, following the relation

$$N(t) = N_0 R_0^{t/\tau} \tag{4.4}$$

where $N(t)$ is the number of infected individuals at time t, N_0 the number at time $t = 0$, R_0 the growth parameter, and τ a characteristic time constant. If $R_0 > 1$, the infection spreads, assuming the underlying system can, in fact, be effectively collapsed into a single point population. In reality, social and geographic dispersion make that impossible, and, indeed, sufficient such dispersion—enough social and economic refugia—can drive a system having an average $R_0 < 1$ into explosive proliferation. The argument is somewhat subtle.

Equation (4.4) can be rewritten for a growth parameter that changes in time as

$$dN/dt = N(t) \left(\frac{\log[R_0(t)]}{\tau} + \frac{t}{\tau} d \log[R_0(t)]/dt \right)$$

$$\equiv N(t) \left(\frac{R^1(t)}{\tau} + \frac{t}{\tau} dR^1/dt \right) \tag{4.5}$$

where the focus now is on the second-order stability of $R^1(t) \equiv \log[R_0(t)]$ in the context of volatility, that is, driven by the dynamics of the stochastic differential equation (Protter 2005)

$$dR_t^1 = f(R_t^1)dt + \sigma R_t^1 dW_t \tag{4.6}$$

Here, σ is the stochastic burden and f a real-valued function having a stable equilibrium value, i.e., $f(R_0^1) = 0$ for some nonzero R_0^1.

It is, however, important to recognize that relations involving R^1 do not have an absorbing state. $R^1 = 0$ simply means that $R_0 = 1$. This has serious implications for any policies based on 'R_0' models.

It is well known (e.g., Appleby et al. 2008) that sufficiently large σ can destabilize Eq. (4.6), driving R^1, and hence R_0, to arbitrarily large values. This can be most easily seen by approximating f near the equilibrium value of R^1 by the linear expression $f \approx \beta - \alpha R^1$, so that $R_0^1 = \beta/\alpha$, and then expanding $(R_t^1)^2$ using the Ito Chain Rule (Protter 2005) to calculate the variance of R^1 from Eq. (4.6). The result is

$$Var(R^1) = \frac{\beta^2}{\alpha - \sigma^2/2} - \frac{\beta^2}{\alpha^2} \tag{4.7}$$

In spite of there being a formal 'equilibrium value' for R^1—here β/α—negative, and thus sufficient to drive infection to zero at infinite time, if $\sigma^2/2 \to \alpha$, R^1 becomes explosively unstable. Enough dispersion, enough stochastic burden, enough social and geographic refugia for the pathogen, and 'R_0' ceases to have meaning. Most particularly, if $R^1 = \log(R_0)$ undergoes an unstable random walk according to Eq. (4.7), then 'R_0' will inevitably be driven into extreme positive magnitudes, generating an exponentially rising infection.

In other words, sufficient political, geographic, or other instabilities over a large enough social landscape can generate an exploding pandemic having virtually complete population penetrance.

In the words of Dr. Gregory Pappas, 'Concentration is not containment'.

4.5 Discussion

A pathogen that evolves the right balance of transmission and virulence can explode from the deepest forest or almost any high intensity farm to the most globalized of population centers. The question of pandemic emergence and spread then becomes entirely dependent on local agroecological structures and processes, even as these are also interpenetrated with global circuits of capital.

Absent the adoption of the strategies and tactics of asymmetric conflict necessary to combat the current lock large-scale agribusiness has on the political processes of virtually all industrialized nations, emergence of a pandemic human analog to African Swine Fever, with a fatality rate of some 50%, seems inevitable. A more complete description of the strategies and tactics of asymmetric conflict can be found in Wallace et al. (2020), and revolves around modes of opposition to colonial power used by the Vietnamese against the Chinese, French, and Americans, and by the Afghans against England, the USSR, and the USA. Roughly analogous strategies and tactics evolved within the Red Army in WWII, culminating in the collapse of Germany's tactically superior Army Group Center in August, 1944.

What is less understood than the risk of a particular pandemic episode, as described in more detail by Wallace and Wallace (2016, Ch.5), is that any single massive loss of life event is only the beginning of the full disaster cascade, leading to a path-dependent trajectory of often much greater morbidity and mortality.

Perhaps the earliest published study of such a possible cascade is the landmark 1962 *New England Journal of Medicine* issue on the health effects of thermonuclear war. The contributors examined continuing loss-of-life in the post-attack period following a 10-weapon, 56 megaton strike on the Boston area. The initial burst and subsequent firestorms were estimated to kill some 2.2 million, but Ervin et al. (1962) describe in detail how the longer-term survival of human populations after such an ecologic upheaval would be precarious, even assuming an intact social structure and the maintenance of a functioning workforce.

Each massive loss-of-life event—pandemic, armed conflict, economic collapse, natural catastrophe—is always followed—often rapidly or in parallel—by its own unique harvest of further social disintegration and death.

The previous chapters detailed how bigotry and corruption fueled COVID-19 death rates within New York City and the NY metropolitan region. This brief chapter summarizes how national and international power relations between Big Agriculture, nation states, and international organizations set the stage for wave after wave of pandemics, some of which will crop huge numbers of lives. To prevent this lethal future scenario these power relations must be cut so that true prevention policies can replace the current ineffective control toolbox.

References

Appleby J, Mao X, Rodkina A (2008) Stabilization and destabilization of nonlinear differential equations by noise. IEEE Trans Autom Control 53:683–691

Bailey N (1975) The mathematical theory of infectious diseases and its applications, 2nd edn. Hafner Press, New York

Ervin F, Glazer J, Aronow S, Nathan D, Coleman R, Avery N, et al. (1962) I. Human and ecologic effects in Massachusetts of an assumed thermonuclear attack on the United States. N Engl J Med 266:1127–1137

Ferguson N et al. (2020) Impact of non-pharmaceutical interventions (NPIs) to reduce COVID-19 mortality and health care demand. Download available from the Imperial College website

Gandolfi A (2013) Percolation methods for SEIR epidemics on graphs. Chapter 2. In Rao V, Durvasula R (eds.) Dynamic models of infectious diseases: volume 2, non vector-borne diseases. Springer, New York, pp 31–58

Parshani R, Carmi S, Havlin S (2010) Epidemic threshold for the susceptible-infectious-susceptible model on random networks. Phys Rev Lett 104:258701

Protter P (2005) Stochastic integration and differential equations: a new approach, 2nd edn. Springer, New York

Shen C, Taleb N, Bar-Yam Y (2020) Review of Ferguson et al. 'Impact of non-pharmaceutical interventions...' Available for download from the New England Complex Systems Institute

Taubenberger J, Morens D (2006) 1918 influenza: the mother of all pandemics. Emerg Infect Dis 12:15–22

Wallace D (1990) In the mouth of the dragon: toxic fires in the age of plastics. Avery, New York

Wallace D, Wallace R (1998) A plague on your houses. Verso, New York

Wallace R, Wallace D, Ullmann JE, Andrews H (1999) Deindustrialization, inner-city decay, and the hierarchical diffusion of AIDS in the USA: how neoliberal and cold war polices magnified the ecological niche for emerging infections and created a national security crisis, Environ. Plan A 31:113–139

Wallace R, Wallace RG (2015) Blowback: new formal perspectives on agriculturally driven pathogen evolution and spread. Epidemiol Infect 143:2068–2080

Wallace R, Liebman A, Bergmann L, Wallace RG (2020) Agribusiness vs. public health: disease control in resource-asymmetric conflict. https://hal.archives-ouvertes.fr/hal-02513883

Wallace RG, Wallace R (eds) (2016) Neoliberal Ebola: modeling disease emergence from finance to forest and farm. Springer, New York

Wu F et al. (2020) A new coronavirus associated with human respiratory disease in China. Nature 579:265–269

Zhou P et al. (2020) A pneumonia outbreak associated with a new coronavirus of probable bat origin. Nature 579:270–273

Chapter 5
Conclusion: Scales of Time in Disasters

5.1 Futures

Chapter 4 explores Ferguson's epidemic model that was the impetus for control of COVID in the United Kingdom. The model failed to include effects of variations in parameters (stochasticity). Furthermore, it and many other models and action plans for control ignore the elephant in the room: disease after disease has jumped from animals to humans. Long co-evolution between natural host and microbe resulted in low virulence to the host and tolerance in the host for relatively benign infection. Sudden large-scale human intrusion into the natural environment, usually by Big Agriculture backed by neo-colonialist capital, enables the jump from animal to humans (Example: Wallace et al. 2016).

In the case of COVID, most scientists involved in the research agree that the virus jumped from bat populations in caves in China and possibly sojourned in an intermediate animal host before the jump to humans. Two possible avenues for the jump are: (1) harvesting of bats for eating and/or traditional Chinese; medicine and (2) changes in agriculture that diminished insect prey of the bats and forced them to forage far afield. New York City and State authorities paid little attention to events in China. Only when cases began emerging in the NY metropolitan region did the disease elicit a response: too late, too clumsy, too biased by class and race/ethnicity.

Wallace (2011) listed the policies that left NYC communities vulnerable to further impacts, with incidence of low-weight births in community districts as the marker of vulnerability. In the pre-WWII era, redlining limited the ability of home-owners and landlords in areas with even a moderate percent of African Americans to get mortgages and home insurance. Krieger et al. (2020) provided a 1938 map of redlining in New York City, showing that the traditional 'ghettoes' were targeted then, including the South Bronx, Harlem, and the poverty belt of Brooklyn from Williamsburg all the way southeast along the border with Queens to Brownsville and East New York. Inability to acquire mortgages and loans blunted the ability

D. Wallace, R. Wallace, *COVID-19 in New York City*, SpringerBriefs in Public
Health, https://doi.org/10.1007/978-3-030-59624-8_5

to renovate and repair housing and was partially responsible for dilapidation in the 'slums'. Thus, 'slums' became synonymous with skin color.

Post-WWII, urban renewal led to massive evictions from the 'slums'. For descriptions of what urban renewal perpetrated in New York City, see Schwartz (1993). For a more national view of urban renewal, see Fullilove (2005). Schwartz (1993) documents the consequences of urban renewal's destruction of social networks and of economic structure of 'slums': loss of social control and support which led to substance abuse, violence, promiscuity, and prostitution, especially in the new public housing. Hinkle and Loring (1977) commented on the paradox of this new clean home environment and the psychological consequences of the loss of community. Urban renewal began the trajectory toward housing famine by removing more low-cost housing than it replaced (see Schwartz 1993 for the numbers of loss and gain).

During pre-civil rights law times, racial steering by realtors was legal and widely practiced (Massey and Denton 1993) so that African Americans and Latinx were segregated into geographic zones which were then targeted for discriminatory policies. After urban renewal became politically toxic and politicians of color began running for citywide offices, the NYC/Rand Institute was created for the ostensible purpose of increasing efficiency in city services, but really to keep communities of color from organizing and voting. Although the Institute developed policies that dealt with issues other than fire service, including weakening rent regulation, the main project turned out to be creation of mathematical models and algorithms that slashed fire control service to poor neighborhoods, especially poor neighborhoods of color. See Wallace and Wallace (1998) for details. These service reductions included changing the number of fire companies responding to alarms and removing fire companies from targeted neighborhoods. During the 1972–1976 period, over a tenth of the fire companies were eliminated. Some of the ones left were transferred to wealthier, whiter neighborhoods (Wallace and Wallace 1998).

Before the slashing, the Fire Department of the City of New York (FDNY) had opened fire companies in neighborhoods of old, overcrowded housing with high fire incidence because of fire service unions' legal action on workload. The increase in firefighting resources leveled off the number and seriousness of the fires. Loss of those new units plus others triggered a fire epidemic and a building abandonment epidemic which destroyed over 200,000 housing units (Wallace and Wallace 1998).

The same communities that had suffered from redlining and urban renewal received a much more literally devastating blow in the form of this planned shrink-age. Planned shrinkage came from the mind of Roger Starr, the real estate industry's prime intellectual who served on the editorial board of *The New York Times* and was briefly NYC Housing Commissioner in the Abraham Beame Administration.

Planned shrinkage implemented Daniel P. Moynihan's Benign Neglect and triaged neighborhoods according to their 'health', a totally subjective judgment based largely on income and race. The 1969 Master Plan called for clearing of several neighborhoods of color, including the South Bronx and Brownsville/East New York, ostensibly to provide land for industry. It planned to forbid the re-renting of units in these neighborhoods when tenants moved out and asserted that minority tenants moved frequently. Thus, whole buildings would be emptied quickly and large areas cleared. Planned shrinkage, not mentioned in the Master Plan, would speed up this process greatly, clearing vast tracts in a few months. The Rand fire deployment models provided a pseudoscientific rationale for planned shrinkage of fire control.

The intertwined epidemics of fire and landlord abandonment of buildings initiated a cascade of consequences described in detail in Wallace and Wallace (1990, 1998, 2011). Their full impacts reached the 24 counties of the NY Metropolitan Region and the national network of American metropolitan regions. We note the extent of the consequences to reveal the intensity and extent of the wound to most neighborhoods of color in New York. Even urban renewal did not trigger a regional and national set of epidemics. The Bronx suffered the widest and deepest wound (see Fig. 3.3). But most of the affected neighborhoods never regained their social, economic, and political functions and had lost the political power to make effective demands for services and resources. They were left open to gentrification, yet another wave of displacement and disempowerment.

Gentrification has speeded the shift in population from poor and dark-skinned to wealthy and white-skinned to the point where Central Harlem (CD 10M) was only 56% Black in 2010 (NYC Planning 2020), whereas it had been 80–90% Black in the 1990s (Infoshare, NYC). The real estate industry has called the South Bronx the new hot spot in its attempt to shift population there. Bushwick which suffered severe damage 1975–1980 is trendy now and gaining a rich Bohemian crowd.

The speed of gentrification depends greatly on tenant harassment by landlords. We lived for 4 years in the neighborhood of Fort George in Washington Heights and witnessed widespread severe harassment of tenants in rent-regulated apartments. Techniques ranged from renting apartments to drug dealers who sold their wares in front of the buildings and held noisy all-night drug parties to failing to make timely repairs to illegally raising rents to buying out old tenants with measly lump sums. Gentrification involved skirting and actual breaking of the housing laws and persistent erosion of the lives of the protected tenants. The city and state made big show of tenant protection but the efforts were just that: show without substance. The proof 'in the pudding' consists of continued tenant harassment.

Under overt and hidden public policies through many decades, segregation within Manhattan and Brooklyn increased. Areas that had been highly mixed (the Upper West Side, Greenwich Village, the Lower East Side, and Washington Heights as Manhattan examples) turned or are turning rich and white. Housing costs spiraled so that 25% or more of the populations of even wealthy areas in the two boroughs suffered with rent stress. Figures 1.1, 2.1, and 2.2 display public health outcomes of the segregation and gross discrimination in Manhattan. If McCord and Freeman (1990) were to repeat their study, the life expectancy differences between Black males in Harlem and white males in white neighborhoods would have increased in the 30 years since their landmark publication, as the maps of premature mortality ranges indicate.

Wallace and Wallace (1998) explored Roger Starr's intellectual smokescreen for clearing land for 'development'. 'Development' sounds good as if useful buildings would replace the useless ones, but it is code for leveling old buildings and putting luxury housing in their place. The UN rapporteur on human rights labeled it 'developmental displacement' and proclaimed it a form of ethnic cleansing (UN Habitat 2011).

Starr (1966, p. 46) stripped poor people of color of their humanity and intrinsic value as human beings:

> Since they have no property, their only marketable asset is hardship in a society pledged to eliminate that hardship which it is unable to ignore. Because this hardship is described to social workers and community organizers who are constitutionally disposed to believe the people they are listening to, and whose luck it is to listen only to the downtrodden and disadvantaged, it seems an immoral suggestion that some of the people displaced by urban renewal might just be exaggerating the sense of deprivation that they feel over their 'lost homes'.

In Starr's view, if one had no property, one had neither value nor morality. This devaluation formed the rationale for Planned Shrinkage. It harkened back to the view of the working and poor classes held during the Industrial Revolution by the robber barons. It was widely held in the FIRE industries. Indeed, one of us (DW) attended a meeting of the NY Federal Reserve and sat among the financial industry representatives who conversed informally before the meeting was called to order. One said to another, 'Can you really believe the low incomes reported by some of these people?' The other simply shook his head in disbelief. These two guys in their $600 suits and $300 shoes imagined hidden wealth and income among 'these people' of the South Bronx, Harlem, and Brownsville. Thus, the poverty of 'these people' could be ignored as unreal and unworthy of relief. Indeed, the phrase 'these people' sorts humans into us and them with 'them' being less than truly human and unworthy of human rights and power.

Application of Planned Shrinkage, a system of triaging essential services to neighborhoods according to their subjectively assigned viability, to fire control required not seeing the inhabitants of the service-deprived neighborhoods as fully human. Otherwise, such deprivation would be murder. Indeed, we found that large numbers of fire deaths were hidden, starting in 1973, the year after the first fire company deprivations. By the time of the discovery and public exposure of the cooked up low numbers, deaths due to the cuts numbered in the hundreds and included many children and elderly. However, indirect deaths due to the cascade of consequences of neighborhood destruction numbered in the tens of thousands at least. If one counts only excess homicides, it comes to about 30,000 cumulatively over years 1978–1993. See Wallace and Wallace (1998) for details. If excess homicides, AIDS mortality, infant mortality, and the other causes of excess deaths due to the massive and sudden loss of housing were added together, the total would probably exceed 100,000 over New York City alone, not counting the ripple into the rest of the metropolitan region described in Chap. 3. Mass murder, largely of people of color, had been committed by public policy. Firefighters also experienced morbidity and mortality directly and indirectly from the toxic workload (Wallace 1982) and despair.

Social epidemiologists who study disparities of morbidity and mortality often ascribe the generation of the conditions that produce these disparities to capitalism and to the class and race/ethnic bigotry on which capitalism depends. However, stripping New York of industry (a blow to capitalism) hints that capitalism isn't the only force at work. It may cloak something older and darker: worship of death and joy in killing. Sometime in the far Greek past, Pluto (god of wealth) became conflated with Hades (god of the underworld).

Every level of government handled the arrival of COVID-19 with extreme negligence. The actions of the Donald Trump Administration actually accelerated the pandemic and led to vastly excessive cases and deaths. We shall not delve into the details of what these actions were and how they played out; the mass media such as *The New York Times, CNN, and the Guardian* have provided analyses. The policies of the Andrew Cuomo Administration had already caused severe public health and safety problems. They reduced hospital and institutional beds to the bone without regard to the needs of the mentally ill or to the possibility of emergencies.

The mentally ill went from institutions to prisons and to short, painful lives, homeless on the streets. Prisons had to create mental health wards. Because of the low number of hospital beds, the Cuomo Administration adopted COVID policies to keep sick people out of hospitals and to limit testing, at first, only to the very ill. People died at home, sometimes not discovered for days. Sick people spread the virus to their family and community. Lack of testing allowed infected individuals with no or few symptoms to spread the virus widely until the rules required use of nose-and-mouth masks and social distancing in public places. During the cresting of the first epidemic wave, Cuomo produced a budget that slashed Medicaid, ensuring that the working poor would avoid contact with medical care providers and would keep working even if ill.

NYC Mayor Bill de Blasio listened to the head of the NYC Health and Hospitals Corporation (HHC) and not to his own Health Commissioner. The head of HHC told him that COVID was nothing serious and that the whole thing would blow over quickly. He also told the Mayor that schools should not be closed. Precious time went by in New York City before the magnitude and utter direness of the pandemic dawned on the Mayor. Even then, he transferred the vital function of contact tracing from NYC Department of Health (DoH) to HHC. DoH had decades of experience in contact tracing and HHC none. Three thousand virgin contact tracers were hired and trained. To their complete surprise, they immediately ran into difficulty getting information from infected individuals, and not simply because of language barriers. The tool deemed most important for curbing COVID spread landed in incompetent hands because of the Mayor's buddy system.

Decades of murderous public policy produced large, dense populations of disempowered, resource-starved people, largely people of color. Homelessness rose during the Michael Bloomberg and de Blasio mayoralties after having been somewhat shaved. Although crime had declined from 1993 on, the prison populations remained high, especially those in and near New York City. Part of this population were the mentally ill who had no institutions to house them and were not supported by community-based treatment and residences. The homeless and prisoners experienced very high COVID incidence and mortality. Homeless shelters and prisons were labeled 'petri dishes'.

Chapter 3 provides the data and analyses of the spread of COVID mortality rates through the 24 counties of the NY metropolitan region. The Bronx became the epicenter of COVID deaths. Manhattan was one of three high-income counties that did not participate in the epidemic system in its early stages. Because of the density of jobs in Manhattan, whether Manhattan is sucked into the epidemic system is important; it would become the new epicenter.

In the face of rising case and death rates, thousands of the wealthy fled the City. Some Manhattan neighborhoods south of 96th Street lost up to 40% of their population. A similar emigration occurred from wealthy Brooklyn neighborhoods. The housing that the wealthy had gobbled up from the working and poor classes, either by demolition/rebuilding or by eviction, emptied out in high proportions. Some went to their second homes in the Hamptons on Long Island or in the counties north of the City in places like the Catskills. Some went further afield to other states. Ironically, many went to states that subsequently wallowed in COVID cases and deaths while New York State and City saw diminishing COVID markers because the bulk of the populace listened to the authorities who told them to stay home, wash their hands, wear masks in public, and socially distance. For details of this exodus, see *The New York Times* (2020).

Despite the grinding of the working class and the poor of color by the authorities and the real estate industry, the great majority of all New Yorkers who did not flee remained as much at home as possible, resisted visiting friends and family, wore masks in public, and practiced social distancing as much as possible. This adherence to public health protection echoes the response of the residents to the little outbreak of smallpox in 1948; with the cooperation of its residents, New York City's health authorities vaccinated several million people in about 10 days (Rosner 1995).

The pandemic and its economic impacts such as massive losses of jobs are predicted by the media to change American society and its economic basis permanently. The NYC real estate industry has begun to feel changes: apartments and houses are on the market now for long periods of time, much longer than before COVID's arrival. Generally, prices have not fallen except for the very highest tier where a $23 million asking price for a mid-Manhattan penthouse may fall 25%. If a significant portion of those who fled decide not to return, matters will definitely take a downward turn for the real estate industry. Manhattan and Brooklyn would bear the brunt of this erosion.

Manhattan and Brooklyn were also the two boroughs with the tightest connections between socioeconomic (SE) factors and between SE factors and public health markers, including premature mortality rate, COVID indicators, and diabetes mortality rate. As was noted in Chap. 2, tight connections indicate brittle, fragile rigidity, a system that cannot adapt with flexibility to an impact. The two boroughs experienced the impacts that the City as a whole did: the disease, the public health requirements for reducing incidence, and the socioeconomic outfalls of the public health requirements. But they also experienced the emigration of large numbers of the rootless wealthy. These impacts all came down suddenly, but the context for their particular shape and intensity required many decades of murderous public policies. Two scales of time thus interacting may end up turning Manhattan and Brooklyn into disaster zones like the Bronx because of the tight connections and rigidity of the systems operating in these boroughs.

Let us consider a possible scenario:

The rootless wealthy flee Manhattan and Brooklyn, never to return. Their peers do not jump in to fill the empty places, and luxury apartments and town houses sit empty. The real estate industry experiences rapid deflation of property values, and taxes paid on properties plummet. The chic stores, restaurants, fitness centers, and other commercial services that arose to fill the desires of the wealthy wither and disappear, leaving empty storefronts which further depress property values. The sales taxes generated by these enterprises also drop, as do the jobs.

With the devaluation of properties, many home-owners and share-holders in cooperatives and condominiums drown underwater: their mortgages exceed the value of their homes. This happened in the 2007–2008 housing crisis and is not an outlandish prediction. Selling at a minimal loss in this market is improbable, and so the homeowners who lost jobs during the pandemic and cannot pay their exorbitant mortgages declare bankruptcy and walk, producing more empty housing units.

Corporations with headquarters in Manhattan find that telecommuting results in productivity equal to or even greater than office-based work. Telecommuting becomes permanent for a high percent of job-holders at a high percent of these corporations which realize immense savings by renting much smaller office space. The real estate industry is hit with another blow and confronts a vast square footage of empty offices in what had been prime rental areas near Wall Street and in midtown Manhattan. Some stores and restaurants that catered to the well-paid financial and corporate professionals must close because their grossly inflated rents cannot be paid anymore.

Cultural institutions such as the Metropolitan Opera, Brooklyn Museum, zoos, and dance companies cannot operate during the repeated lockdowns from waves of the pandemic. They furlough their employees and artists, keeping only a minimal staff for security and collection conservation. The zoos minimize their collections and try to ride out the extended pandemic on a skeleton staff of keepers and veterinarians.

Unemployment rate rises for all SE classes from the professionals in the real estate industry and in the establishments that could not grapple with the repeated lockdowns from waves of the pandemic to the cleaning personnel who worked at the now-shuttered offices, stores, and apartment buildings.

State and city budgets collapse under the twin burdens of lower tax revenues and higher expenditures for unemployment benefits and other emergencies, including for hospitals facing the repeated waves of pandemic coming in from the states that mismanaged the early waves. The campaign funds of elected officials aren't receiving the usual donations from the FIRE industries. The reduced numbers of the rootless wealthy in places like the Upper West Side and Park Slope reduce the power of these areas. The helpless cultural institutions remain empty. City agencies begin assessing where to make deep cuts in service. The revival and change of mode of operating of the civil rights movement puts limitations on how much poor neighborhoods of color could suffer from further service reductions from the already inadequate levels meted out to them. Semi-empty glitzy neighborhoods, the Lincoln Center area, Wall Street, and other places that had previously hogged city services receive the big cuts which are justified on the basis of reduced population, reduced tax revenues, and ghost town status of the cultural institutions. Garbage collection and street cleaning, funding of schools and libraries, housing inspections, street pavement repairs, policing, health department inspection of food services (restaurants and grocery stores), bus service, and (yes) fire prevention and control resources—all are slashed with concentration of the reductions in the formerly gentrified neighborhoods of the rootless nouveau riche or wannabe riche.

High rates of unemployment impose communal desperation, and families and individuals organize. They squat in empty housing and establish cooperative food distribution centers by buying directly in quantity from farming cooperatives. Families whose grandparents had suffered eviction from the present cultural and luxury areas re-inhabit them. These areas come to resemble Diocletian's Palace, the Alhambra and the Coliseum in the eighteenth and nineteenth centuries when the homeless poor and refugees lived there.

Perhaps this scenario is over-the-top and would never come even close to reality. We don't know how this pandemic and its demographic and socioeconomic cascade will evolve. We do know that Manhattan and Brooklyn lack resilience and the flexibility to adapt. Further impacts lead to regime change, as Holling (1973) terms it, in such brittle, rigid systems. Deep impacts rendered on a short timescale interact badly with systems that embrittled over a long timescale, like a fire racing through a forest with dense, long-accumulated dead wood. We also know that even if we manage to survive and thrive after COVID-19 dies down, other viruses wait in the wings to be loosed on human populations as Big Agriculture and the extractive industries pillage the earth.

> 'My name is Ozymandias, king of kings:
> Look on my works, ye Mighty, and despair!'
> Nothing beside remains. Round the decay
> Of that colossal Wreck, boundless and bare
> The lone and level sands stretch far away.

> — *Ozymandias*, Percy Bysshe Shelley, 1818

References

Fullilove M (2005) Root shock: how tearing up city neighborhoods hurts America and what we can do about it. Ballantine Books, New York

Hinkle L, Loring W (1977) The effects of the man-made environment on health and behavior. DHEW. Publication No. (CDC)77-8318. Center for Disease Control, Washington DC

Holling CS (1973) Resilience and stability of ecosystems. Ann Rev. Ecol Syst 4:1–23

Infoshare NYC (2020) Accessed 5 June, 2020. Data from US Censuses 1980 and 1990

Krieger N et al (2020) Structural racism, historical redlining, and risk of preterm birth in New York City, 2013–2017. Am J Public Health 110:1046–1053

Massey D, Denton N (1993) American Apartheid: segregation and the making of the underclass. Harvard University Press, Cambridge

McCord C, Freeman H (1990) Excess mortality in Harlem. N Engl J Med 322:173–177

NYC Dept. of Planning (2020) Accessed 2020. https://communityprofiles.planning.nyc.gov

NY Times (2020) https://www.nytimes.com/interactive/2020/05/16/nyregion/nyc-coronavirus-moving-leaving.html

Rosner D (1995) Hives of sickness. Ritgers University Press, New Brunswick

Schwartz J (1993) The New York approach: Robert Moses, urban liberals, and redevelopment of the inner city. Ohio State University Press, Columbus

Starr R (1966) Urban choices: the city and its critics. Pelican Books, New York and London

Habitat UN (2011) United nations human rights. Office of the High Commissioner. Losing your home: assessing the impact of eviction. UNON Publishing Services Section, Nairobi

Wallace D (2011) Discriminatory mass de-housing and low-weight births: scales of geography, time, and level. J Urban Health 88:454–468

Wallace D, Wallace R (1998) A plague on your houses: how New York City was burned down and National Public Health crumbled. Verso, New York and London

Wallace D, Wallace R (2011) Consequences of massive housing destruction: the New York City fire epidemic. Build Res Inf 39:395–411

Wallace R (1982) The New York City fire epidemic as a toxic phenomenon. Int Arch Occup Environ Health 50:33–51

Wallace R, Bergman L, Hogerwerf L, Koch R, Wallace RG (2016) Neoliberal Ebola: modeling disease emergence from finance to forest and farm. Springer, Cham

Wallace R, Wallace D (1990) Origins of public health collapse in New York City: the dynamics of planned shrinkage, contagious urban decay, and social disintegration. Bull NY Acad Med 66:391–434

Correction to: COVID-19 in New York City

Correction to:
D. Wallace, R. Wallace, *COVID-19 in New York City*,
SpringerBriefs in Public Health,
https://doi.org/10.1007/978-3-030-59624-8

The original version of this book was inadvertently published with incorrect captions for Fig. 1.1 of Chapter 1 and Fig. 2.1 and Fig. 2.2 of Chapter 2. It also had an incorrect Fig. 3.1 in Chapter 3. This has now been corrected and an erratum to this book can be found at https://doi.org/10.1007/978-3-030-59624-8_6.

The updated online version of these chapters can be found at
https://doi.org/10.1007/978-3-030-59624-8_1
https://doi.org/10.1007/978-3-030-59624-8_2
https://doi.org/10.1007/978-3-030-59624-8_3

Index

Printed in the United States
By Bookmasters